SpringerBriefs in Computer Science

Series editors

Stan Zdonik, Brown University, Providence, Rhode Island, USA
Shashi Shekhar, University of Minnesota, Minneapolis, Minnesota, USA
Xindong Wu, University of Vermont, Burlington, Vermont, USA
Lakhmi C. Jain, University of South Australia, Adelaide, South Australia, Australia
David Padua, University of Illinois Urbana-Champaign, Urbana, Illinois, USA
Xuemin Sherman Shen, University of Waterloo, Waterloo, Ontario, Canada
Borko Furht, Florida Atlantic University, Boca Raton, Florida, USA
V. S. Subrahmanian, University of Maryland, College Park, Maryland, USA
Martial Hebert, Carnegie Mellon University, Pittsburgh, Pennsylvania, USA
Katsushi Ikeuchi, University of Tokyo, Tokyo, Japan
Bruno Siciliano, Università di Napoli Federico II, Napoli, Italy
Sushil Jajodia, George Mason University, Fairfax, Virginia, USA
Newton Lee, Institute for Education, Research, and Scholarships in Los Angeles, California

SpringerBriefs present concise summaries of cutting-edge research and practical applications across a wide spectrum of fields. Featuring compact volumes of 50 to 125 pages, the series covers a range of content from professional to academic. Typical topics might include:

- A timely report of state-of-the art analytical techniques
- A bridge between new research results, as published in journal articles, and a contextual literature review
- A snapshot of a hot or emerging topic
- An in-depth case study or clinical example
- A presentation of core concepts that students must understand in order to make independent contributions

Briefs allow authors to present their ideas and readers to absorb them with minimal time investment. Briefs will be published as part of Springer's eBook collection, with millions of users worldwide. In addition, Briefs will be available for individual print and electronic purchase. Briefs are characterized by fast, global electronic dissemination, standard publishing contracts, easy-to-use manuscript preparation and formatting guidelines, and expedited production schedules. We aim for publication 8–12 weeks after acceptance. Both solicited and unsolicited manuscripts are considered for publication in this series.

More information about this series at http://www.springer.com/series/10028

Jie Cao • Quan Zhang • Weisong Shi

Edge Computing: A Primer

 Springer

Jie Cao
Wayne State University
Detroit, MI, USA

Quan Zhang
Wayne State University
Detroit, MI, USA

Weisong Shi
Wayne State University
Detroit, MI, USA

ISSN 2191-5768   ISSN 2191-5776   (electronic)
SpringerBriefs in Computer Science
ISBN 978-3-030-02082-8   ISBN 978-3-030-02083-5   (eBook)
https://doi.org/10.1007/978-3-030-02083-5

Library of Congress Control Number: 2018958959

This Springer imprint is published by the registered company Springer Nature Switzerland AG
The registered company address is: Gewerbestrasse 11, 6330 Cham, Switzerland

Contents

Chapter 1
Introduction

The proliferation of the Internet of Things and the success of rich cloud services have pushed the horizon of a new computing paradigm, Edge computing, which calls for processing the data at the edge of the network. Edge computing has the potential to address the concerns of response time requirement, battery life constraint, bandwidth cost saving, as well as the data safety and privacy. In this book, we introduce the definition of Edge computing, followed by several case studies, ranging from cloud offloading to smart home and city, as well as collaborative Edge to materialize the concept of Edge computing. Finally, we present several challenges and opportunities in the field of Edge computing and hope this book will gain attention from the community and inspire more research in this direction.

Cloud computing has tremendously changed the way we live, work, and study since its inception around 2005 [1]. For example, Software as a Service (SaaS) instances, such as Google Apps, Twitter, Facebook, and Flickr, have been widely used in our daily life. Moreover, scalable infrastructures, as well as processing engines developed to support cloud service, are also significantly influencing the way of running the business, for instance, Google File System [2], MapReduce [3], Apache Hadoop [4], Apache Spark [5], and so on.

Internet of Things (IoT) was first introduced to the community in 1999 for supply chain management [6], and then the concept of "making a computer sense information without the aid of human intervention" was widely adapted to other fields such as healthcare, home, environment, and transports [7, 8]. Now with IoT, we will arrive in the post-Cloud era, where there will be a significant quality of data generated by things that are immersed in our daily life, and many applications will also be deployed at the edge to consume these data. By 2019, data produced by people, machines, and things will reach 500 zettabytes, as estimated by Cisco

Global Cloud Index, however, the global data center IP traffic will only reach 10.4 zettabytes by that time [9]. By 2019, 45% of IoT-Created data will be stored, processed, analyzed, and acted upon close to, or at the Edge of, the network [10]. There will be 50 billion things connected to the Internet by 2020, as predicted by Cisco Internet Business Solutions Group [11]. Some IoT applications might require short response time, some might involve private data, and some might produce a large quantity of data which could be a heavy load for networks. Cloud computing is not efficient enough to support these applications.

1.1 What Is Edge Computing

Data is increasingly produced at the edge of the network. Therefore, it would be more efficient also to process the data at the edge of the network. Previous work such as micro DataCenter [12, 13], Cloudlet [14], and fog computing [15] has been introduced to the community because Cloud computing is not always efficient for data processing when the data is produced at the edge of the network. In this section, we list some reasons why Edge computing is more efficient than Cloud computing for some computing services, and then we give our definition and understanding of Edge computing.

1.1.1 Why Do We Need Edge Computing

Push from Cloud Services

Putting all the computing tasks on the cloud has been proved to be an efficient way for data processing since the computing power on the cloud outclasses the capability of the things at the edge. However, compared to the fast developing data processing speed, the bandwidth of the network has come to a standstill. With the growing quantity of data generated at the edge, the speed of data transportation is becoming the bottleneck for the Cloud-based computing paradigm. For example, about 5 Gigabyte data will be generated by a Boeing 787 every second [16], but the bandwidth between the airplane and either satellite or base station on the ground is not large enough for data transmission. Consider an autonomous vehicle as another example. 1 Gigabyte data will be generated by the car every second, and it requires real-time processing for the vehicle to make correct decisions [17]. If all the data needs to be sent to the cloud for processing, the response time would be too long. Not to mention that current network bandwidth and reliability would be challenged for its capability of supporting a large number of vehicles in one area. In this case, the data needs to be processed at the edge for shorter response time, more efficient processing and smaller network pressure.

Pull from the Internet of Things

Almost all kinds of electrical devices will become part of IoT, and they will play
the role of data producers as well as consumers, such as air quality sensors, LED
bars, streetlights and even an Internet-connected microwave oven. It is safe to infer
that the number of things at the Edge of the network will develop to more than
billions in a few years. Thus, the raw data produced by them will be enormous,
making traditional Cloud computing not efficient enough to handle all these data.
This means most of the data produced by IoT will never be transmitted to the cloud.
Instead, it will be consumed at the edge of the network.

Fig. 1.1   Cloud computing paradigm

Figure 1.1 shows the conventional Cloud computing structure. Data producers
generate raw data and transfer it to cloud, and data consumers send a request
for consuming data to the cloud, as noted by the solid blue line. The red dotted
line indicates the request for consuming data being sent from data consumers to
cloud, and the green represents the result from the cloud-dotted line. However, this
structure is not sufficient for IoT. Firstly, data quantity at the edge is too large, which
will lead to huge unnecessary bandwidth and computing resource usage. Secondly,
the privacy protection requirement will pose an obstacle for Cloud computing in
IoT. Lastly, most of the end nodes in IoT are energy constrained things, and the
wireless communication module is usually very energy hungry, so offloading some
computing tasks to the edge could be more energy efficient.

Change from a Data Consumer to Producer

In the Cloud computing paradigm, the end devices at the edge usually play as a data
consumer, for example, watching a YouTube video on your smartphone. However,
people are also producing data nowadays on their mobile devices. The change from
a data consumer to data producer/consumer requires more function placement at
the edge. For example, it is very normal that people today take photos or do video
recording then share the data through a cloud service such as YouTube, Facebook,
Twitter or Instagram. Moreover, every single minute, YouTube users upload 72 h
of new video content; Facebook users share nearly 2.5 million pieces of content;
Twitter users tweet nearly 300,000 times; Instagram users post nearly 220,000 new

photos [18]. However, the image or video clip could be reasonably large, and it would occupy much bandwidth for uploading. In this case, the video clip should be demised and adjusted to proper resolution at the edge before uploading to the cloud. Another example would be wearable health devices. Since the physical data collected by the things at the Edge of the network is usually private, processing the data at the edge could protect user privacy better than uploading raw data to the cloud.

1.1.2 Key Techniques that Enable Edge Computing

VMs and Containers

VMs have served Cloud computing very well in the past. Inherited from VMs, containers can be running directly on top of the physical infrastructure and offer virtualization on OS level.

Due to the design of shared OS, the size of the containers can be constrained to MB level, and it might only take several seconds as startup time. The light of the containers fits Edge computing applications very well since the resources requirements are usually limited such as storage size and response time.

Software Defined Networking (SDN)

Edge computing pushes the computational infrastructure to the proximity of the data source, and the computing complexity will also increase correspondingly.

SDN provides a cost-effective solution for Edge network virtualization and simplifies the network complexity by offering the automatic Edge device reconfiguration and bandwidth allocation. Edge devices could be set up and deployed in a plug-and-play manner enabled by SDN. Also, SDN is a promising solution for Edge system security such as IoT, smart home, and smart city.

Content Delivery/Distribution Network (CDN)

CDN is not proposed for Edge computing originally. However, the concept of caching the content to the Edge servers near the data consumers matches the rationale of Edge computing very well.

As the upstream server that delivers the content is becoming the bottleneck of the web due to the increasing web traffic, CDN can offer data caching at the Edge of the network with scalability and save both the bandwidth cost and page load time significantly.

Cloudlets and Micro Data Centers (MDC)

Cloudlets and Microdata centers are the small-scale cloud data centers with mobility enhancement. They can be used as the gateway between Edge/mobile devices and the cloud. The computing power on the Cloudlets or MDCs could be accessed with lower latency by the Edge devices due to the geographical proximity. Essential computing tasks for Edge computing such as speech recognition, language processing, machine learning, image processing, and augmented reality could be deployed on the Cloudlets or MDCs to reduce the resource cost.

1.1.3   Edge Computing Definition

Edge computing refers to the enabling technologies allowing computation to be performed at the edge of the network, on downstream data on behalf of cloud services and upstream data on behalf of IoT services. Here we define "Edge" as any computing and network resources along the path between data sources and cloud data centers. For example, a smartphone is an edge between body things and cloud, a gateway in a smart home is the edge between home things and cloud, a Micro Data Center (MDC) and a Cloudlet [14] is the edge between a mobile device and cloud. The rationale of Edge computing is that ***computing should happen at the proximity of data sources***. From our point of view, Edge computing is interchangeable with Fog computing [19], but Edge computing focus more on the Things side, while Fog computing focuses more on the infrastructure side. We envision that Edge computing could have as big an impact on our society as has the Cloud computing.

Figure 1.2 illustrates the two-way computing streams in Edge computing. In the Edge computing paradigm, the things not only are data consumers but also play as data producers. At the edge, the things cannot only request service and content from the cloud but also perform the computing tasks from the cloud. Edge can perform computing offloading, data storage, caching and processing, as well as distribute request and delivery service from cloud to user. With those jobs in the network, the edge itself needs to be well designed to meet the requirement efficiently in services such as reliability, security, and privacy protection.

1.1.4   Edge Computing Benefits

In Edge Computing we want to put the computing at the proximity of data sources. This has several benefits compared to traditional Cloud-based computing paradigm. Here we use several early results from the community to demonstrate the

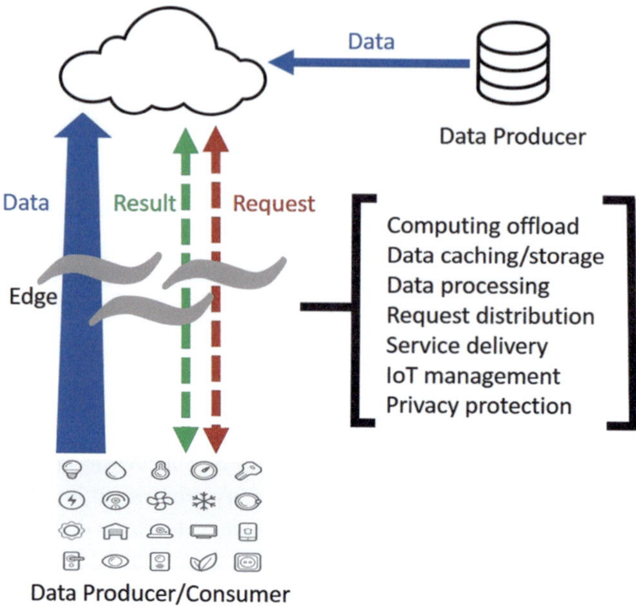

Fig. 1.2 Edge computing paradigm

potential benefits. Researchers built a proof-of-concept platform to face recognition application in [20], and the response time is reduced from 900 to 169 ms by moving computation from the cloud to the Edge. In [21], the researchers use Cloudlets to offload computing tasks for wearable cognitive assistance, and the result shows that the improvement of response time is between 80 to 200 ms. Moreover, the energy consumption could also be reduced by 30–40% by cloudlet offloading. CloneCloud in [22] combine partitioning, migration with merging, and on-demand instantiation of partitioning between mobile and the cloud, and their prototype could reduce 20x running time and energy for tested applications.

1.1.5 Edge Computing Systems

Several open source systems for Edge computing have been developed and deployed. Here we list some and hope future Edge computing practitioners could benefit from them.

Apache Edgent [23] is a programming model that can process streams of data locally at the Edge devices or gateways in real-time. Data is determined by Edgent to be stored or analysis at Edge device or back-end systems. Edgent enables the application to transform to sending only essential and useful data to the server from sending the continuous raw data flow. By doing this, the amount of data that are transmitted and stored to the server could be significantly reduced.

OpenStack [24] is a cloud operating system which use a data-center to control compute, storage, and networking resources. It also provides management tools through a dashboard, as well as a web interface to the users. The fundamental infrastructure of OpenStack can be deployed at Edge devices, and the distributed software of OpenStack provide support for virtual machines and container technologies, which are vital technologies enable Edge computing.

EdgeX Foundry [25] is a vendor-neutral open interop platform for the IoT and Edge computing. It is an interoperability framework hosted by the Linux Foundation within a full hardware/OS platform. The interested parties of Edge computing can collaborate on IoT solutions freely using current communication standards with their proprietary innovations. EdgeX Foundry focuses on Industrial IoT by leveraging both cloud principles and specific needs of IoT communication protocols. It can also be scaled down to Edge devices and provides security and system management for Edge nodes.

There are more open source solutions for Edge computing available to the community such as machine.io [26] and IOTRACKS [27], and the Edge computing practitioners can explore the solutions that best fit their applications and systems.

Along with the open source solutions, there are also business systems which enable advanced data analytics as well as artificial intelligence at the edge of the network from cloud service providers such as Azure IoT Edge from Microsoft, Google Cloud IoT, and AWS Greengrass from Amazon, etc.

Azure IoT Edge [28] moves the data analytics from the cloud to the Edge devices. Three components together make up the Azure IoT Edge including the IoT Edge modules, the IoT Edge runtime, and the cloud-based interface. The IoT Edge runtime enables the cloud logic on Edge devices to manage the communications and operations. Meanwhile, multiple IoT Edge modules can be running on the Edge device as Docker compatible containers to perform Azure services, third-party services, or customized code. The IoT Edge cloud interface allows the users to distribute and monitor the workloads on the Edge devices.

Amazon AWS Greengrass [29] is the software that enables local computer, communication, data caching, sync, and analytics on the Edge device. The Greengrass core is the runtime that enables the local execution of AWS Lambda, messaging, device shadows, and security.

1.2 Overview of the Book

With the push from cloud services and pull from IoT, we envision that the edge of the network is changing from data consumer to data producer as well as data consumer. In this book, we attempt to contribute to the concept of Edge computing. We start from the analysis of why we need Edge computing, and then we give our definition and vision of Edge computing.

Followed by definition, to further explain Edge computing in a detailed manner, we will move on present four case studies for Edge computing.

The first case study will be a home operating system for Edge computing. In Chap. 2, we will present our vision of a home operating system and introduce how the system is designed to have a stable programming interface and a self-management capability. We will also discuss how the data should be abstracted, stored, and evaluated in the smart home.

In Chap. 3, we will introduce the second case study as a framework for hybrid Cloud-Edge analytics. This framework fuses data from multiple stakeholders as a virtually shared dataset that is a collection of data and predefined functions by data owners. Application on this framework will be breaking down into subservices so that a user can directly subscribe to intermediate data and compose new applications by leveraging existing sub-services. An easy-to-use programming interface is also provided in the framework for both service providers and end users.

The third case study will be video analytics on Edge computing platform. Video analytics is becoming more and more important for applications in public safety, counter-terrorism, self-driving cars, VR/AR, etc. In Chap. 4, we will investigate providing video analytic services to latency-sensitive applications in Edge computing environment. This case collaborates with the nearby client, edge, and remote cloud nodes, and transfers video feeds into semantic information at places closer to the users in early stages.

After present the case studies, we list some challenges and opportunities in Edge computing in Chap. 5, such as programmability, naming, data abstraction, service management, privacy, and security, as well as optimization metrics that are worth future research and study.

Then we will introduce existing Edge computing tools and several case studies that Edge computing is practically implemented in Chap. 6. The tools and software appear in this chapter are a tip of the thousands of open-sourced or production-ready tools and software that available in the community. Thus, this chapter serves as a high-level literature review of representatives of the most popular tools and software.

References

1. M. Armbrust, A. Fox, R. Griffith, A. D. Joseph, R. Katz, A. Konwinski, G. Lee, D. Patterson, A. Rabkin, I. Stoica *et al.*, "A view of cloud computing," *Communications of the ACM*, vol. 53, no. 4, pp. 50–58, 2010.
2. S. Ghemawat, H. Gobioff, and S.-T. Leung, "The google file system," in *ACM SIGOPS operating systems review*, vol. 37, no. 5. ACM, 2003, pp. 29–43.
3. J. Dean and S. Ghemawat, "Mapreduce: simplified data processing on large clusters," *Communications of the ACM*, vol. 51, no. 1, pp. 107–113, 2008.
4. K. Shvachko, H. Kuang, S. Radia, and R. Chansler, "The hadoop distributed file system," in *Mass Storage Systems and Technologies (MSST), 2010 IEEE 26th Symposium on*. IEEE, 2010, pp. 1–10.

5. M. Zaharia, M. Chowdhury, M. J. Franklin, S. Shenker, and I. Stoica, "Spark: cluster computing with working sets," in *Proceedings of the 2nd USENIX conference on Hot topics in cloud computing*, vol. 10, 2010, p. 10.
6. K. Ashton, "That 'internet of thing' thing," *RFiD Journal*, vol. 22, no. 7, pp. 97–114, 2009.
7. H. Sundmaeker, P. Guillemin, P. Friess, and S. Woelfflé, "Vision and challenges for realising the internet of things," 2010.
8. J. Gubbi, R. Buyya, S. Marusic, and M. Palaniswami, "Internet of things (iot): A vision, architectural elements, and future directions," *Future Generation Computer Systems*, vol. 29, no. 7, pp. 1645–1660, 2013.
9. "Cisco global cloud index: Forecast and methodology, 2014–2019 white paper," 2014.
10. "IDC futurescape: Worldwide internet of things 2016 predictions," 2015.
11. D. Evans, "The internet of things: How the next evolution of the internet is changing everything," *CISCO white paper*, vol. 1, p. 14, 2011.
12. A. Greenberg, J. Hamilton, D. A. Maltz, and P. Patel, "The cost of a cloud: research problems in data center networks," *ACM SIGCOMM computer communication review*, vol. 39, no. 1, pp. 68–73, 2008.
13. E. Cuervo, A. Balasubramanian, D.-k. Cho, A. Wolman, S. Saroiu, R. Chandra, and P. Bahl, "Maui: making smartphones last longer with code offload," in *Proceedings of the 8th international conference on Mobile systems, applications, and services*. ACM, 2010, pp. 49–62.
14. M. Satyanarayanan, P. Bahl, R. Caceres, and N. Davies, "The case for vm-based cloudlets in mobile computing," *Pervasive Computing, IEEE*, vol. 8, no. 4, pp. 14–23, 2009.
15. F. Bonomi, R. Milito, J. Zhu, and S. Addepalli, "Fog computing and its role in the internet of things," in *Proceedings of the first edition of the MCC workshop on Mobile cloud computing*. ACM, 2012, pp. 13–16.
16. "Boeing 787s to create half a terabyte of data per flight, says virgin atlantic," https://datafloq.com/read/self-driving-cars-create-2-petabytes-data-annually/172.
17. "Self-driving cars will create 2 petabytes of data, what are the big data opportunities for the car industry?" http://www.computerworlduk.com/news/data/boeing-787s-create-half-terabyte-of-data-per-flight-says-virgin-atlantic-3433595/.
18. "Data never sleeps 2.0," https://www.domo.com/blog/2014/04/data-never-sleeps-2-0/.
19. "Openfog architecture overview," *OpenFog Consortium Architecture Working Group*, 2016.
20. S. Yi, Z. Hao, Z. Qin, and Q. Li, "Fog computing: Platform and applications," in *Hot Topics in Web Systems and Technologies (HotWeb), 2015 Third IEEE Workshop on*. IEEE, 2015, pp. 73–78.
21. K. Ha, Z. Chen, W. Hu, W. Richter, P. Pillai, and M. Satyanarayanan, "Towards wearable cognitive assistance," in *Proceedings of the 12th annual international conference on Mobile systems, applications, and services*. ACM, 2014, pp. 68–81.
22. B.-G. Chun, S. Ihm, P. Maniatis, M. Naik, and A. Patti, "Clonecloud: elastic execution between mobile device and cloud," in *Proceedings of the sixth conference on Computer systems*. ACM, 2011, pp. 301–314.
23. "Apache edgent," http://edgent.apache.org/.
24. "Openstack," https://www.openstack.org/edge-computing/.
25. "Edgex foundry," https://www.edgexfoundry.org/.
26. "macchina.io," https://macchina.io//.
27. "Iotracks," https://iotracks.com/.
28. "Azure iot edge," https://azure.microsoft.com/en-in/services/iot-edge/.
29. "Aws greengrass," https://aws.amazon.com/greengrass/.

Chapter 2
EdgeOS$_H$: A Home Operating System for Internet of Everything

Smart home as a typical IoE application is widely adopted in people's lives. Edge Computing has the potential to empower the smart home, but it needs more contribution from the community before it genuinely benefits our lives. In this chapter, we present the vision of EdgeOS$_H$, a home operating system for the Internet of Everything. We further discuss practical challenges, namely programming interface, self-management, data management, security & privacy, and naming, as well as non-functional challenges, such as user experience, system cost, delay, and the lack of availability of open testbed. Within each challenge, we also discuss the potential directions that are worth further investigation.

2.1 Introduction

Edge Computing [1, 2] could be a good solution for the smart home operating system by allowing computation to be performed at the edge of the network, on downstream data on behalf of Cloud services and upstream data on behalf of IoE devices. We observed that with more and more connected things becoming available at home, many people are doing DIY style smart home design and installation. The lack of a home operating system makes it very difficult to manage devices, data, and services. This is due to most of the systems nowadays work in a silo-based manner and can not be connected or communicate with other systems, as shown in Fig. 2.1. To solve this problem using Edge Computing, we introduce the concept of EdgeOS$_H$,[1] which is a home operating system for the Internet of Everything.

---

[1]EdgeOS stands for the operating system for Edge Computing. For different environments, EdgeOS could have multiple variations. For example, EdgeOS$_H$ for smart home, EdgeOS$_V$ for vehicle, EdgeOS$_B$ for smart building, and so on.

J. Cao et al., *Edge Computing: A Primer*, SpringerBriefs in Computer Science,
https://doi.org/10.1007/978-3-030-02083-5_2

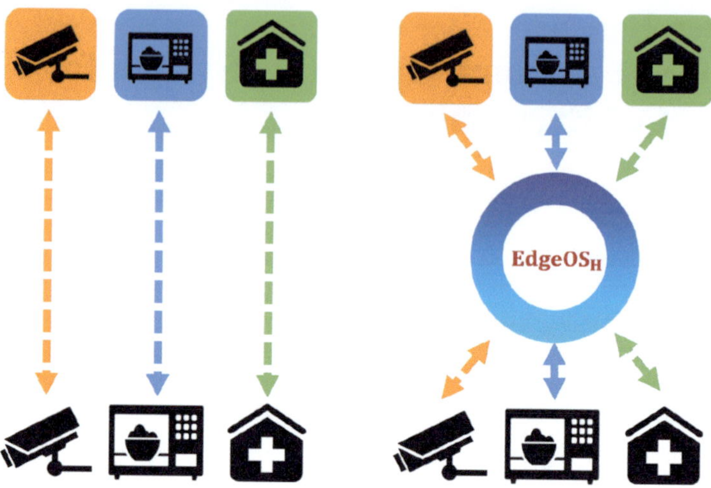

Fig. 2.1 A comparison of silo-based vs. EdgeOS-based smart home

With Edge Computing, the devices and services in the home could be connected to a central EdgeOS. This paradigm manages the devices and services more efficiently and easily.

2.2   Related Work

Smart home has drawn much attention from the community since ubiquitous computing became popular. Researchers and practitioners spare no effort in applying novel ideas as well as technologies in the domestic environment. In 2002 [3] the research team from MIT has developed a living laboratory using context-aware sensing to empower the home. In their understanding, a smart home will be a place where people can live a healthier and longer life via the help of digital and robotic agents. Resource consumption will be reduced in the smart home, and the home will be fully automated where occupants do not need to think about daily tasks at all. Kientz et al. from Georgia Tech developed Aware Home [4] in 1998, which is viewed as the "glimpse of the most advanced domestic technologies for the potential future home."

With the proliferation of the high-speed Internet and the Internet of Everything, more and more products for the smart home are also available on the market. A smart device such as iRobot, Philips Hue, and Nest learning thermostat, etc. shows that homeowners are ready to embrace smart devices in their daily lives. Amazon Echo, Samsung SmartThings, and Google Home provide a hub and user interface for occupants to interact with connected devices. HomeOS from Microsoft and

HomeKit from Apple enable a framework for communicating with and controlling connected accessories in a smart home.

Our understanding of smart home is that it should be an *automated* and *energy efficient* domestic place where occupants could enjoy *a healthier and more comfortable life*. A smart home should come with *self-awareness, self-management, and self-learning* ability to satisfy and improve occupants' lifestyles.

Self-awareness means the home should be available to sense the occupants' status and domestic data automatically. How many people are in the home? Where are they? Are they sleeping? Status information of occupants like these should be derived from the domestic data such as room temperature, motion sensor, camera, etc. Self-awareness is a necessary capability for self-management and self-learning, and there is some ongoing research in this field [5–10].

As a smart home is supposed to arrange everything automatically and free occupants' hands, self-management plays a crucial role in the whole system [11].

Self-learning refers to the ability to profile the occupant's behavior based on historical data to make personalized configuration of the home. Researchers and practitioners have contributed some work to provide this ability to the home environment [12–15].

Despite the research and smart home products being readily available, some challenges always exist before we have a real smart home. In [16], Edwards and Grinter listed seven challenges that should be addressed for the smart home, including the accidentally domestic, impromptu interoperability, the lack of system administrator, adoption of domestic technologies, social implications, the reliability of the smart home, and ambiguity in ubiquitous sensing and computing. Sixteen years later, these challenges remain unsolved in today's home environment. When Mennicken et al. examined the recent research with the observation of industry in [17], they provided three challenges in the current smart home such as meaningful technologies, complex domestic spaces, and human-home collaboration.

Although challenges and potential solutions have been examined in previous research, we think that there are still some issues that are worth raising about smart home. In the next section, we will introduce the overview and structure of EdgeOS$_H$, and we hope to contribute to the consensus among various disciplines that make up the smart home.

2.3 EdgeOS$_H$: Overview and Design

In a smart home, data is produced by various sensors and devices in the domestic place. Moreover, the data is also consumed at home to serve the occupants. With the rational *"Computing should happen at the proximity of data sources"* [1], we think that the idea of Edge Computing fits perfectly and should be deployed as the computing paradigm for the smart home. In order to apply Edge Computing to the smart home, we propose EdgeOS$_H$, which is a smart home operating system for the

Internet of Everything, as shown in Fig. 2.2. EdgeOS$_H$ is the bridge to connect the devices at home with the Cloud, home occupants, and developers. For the Cloud, EdgeOS$_H$ can upstream/downstream data and computing requests on behalf of the devices. For home occupants, EdgeOS$_H$ provides collaboration between humans and home. For service practitioners, EdgeOS$_H$ is capable of reducing the complexity of development by offering a unified programming interface. For the smart home, EdgeOS$_H$ is the brain that manages the data, devices, and services while guaranteeing the security and privacy of the data.

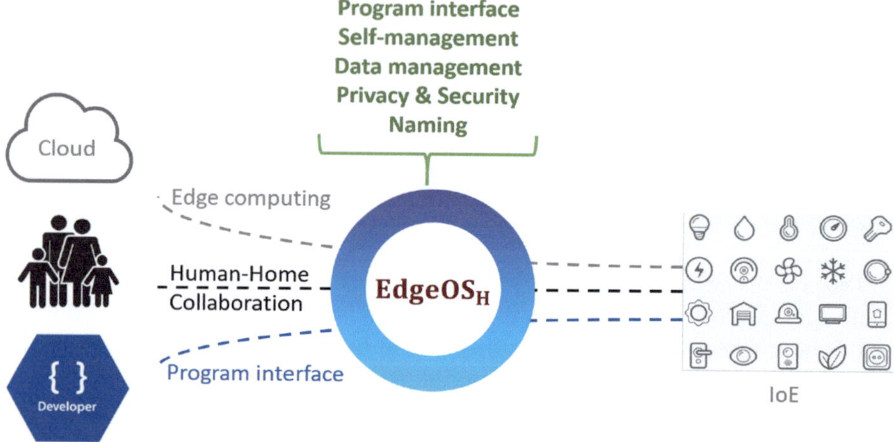

Fig. 2.2 The overview of EdgeOS$_H$

Applying Edge Computing in the smart home will bring several benefits. First, the network load could be reduced if the data is processed at home rather than uploaded to the Cloud. This is important for the domestic environment considering the bandwidth is usually limited. Second, service response time could be decreased since the computing takes place closer to both data producer and consumer. Third, the data could be better protected from an outside attacker since most of the raw data will never go out of the home.

Compared to conventional computing platforms such as PCs, smartphones, and cloud, smart home has its specific characteristics.

First, for PCs or smartphones, the operating system can efficiently manage all the hardware resources since the manufacturer fixed design limits them. However, the home environment is very dynamic, which means the home operating system will face various hardware provided by different manufacturers. The dynamic environment brings about new challenges in communication and management. Moreover, the current smart home applications usually work in a silo-based manner rather than attached to a specific operating system. Therefore, how EdgeOS$_H$ can manage various combinations of devices and services is still a considerable challenge.

Second, traditional operating systems are usually resource-oriented. In the conventional computing platform, e.g., laptops and mobile devices, the most critical responsibility of the operating system is resource management. However, the smart home is a data-oriented environment, which means services in a smart home should interact directly with the data collected by the devices, rather than the resources, or in another word, the specific devices.

2.3.1 Overview

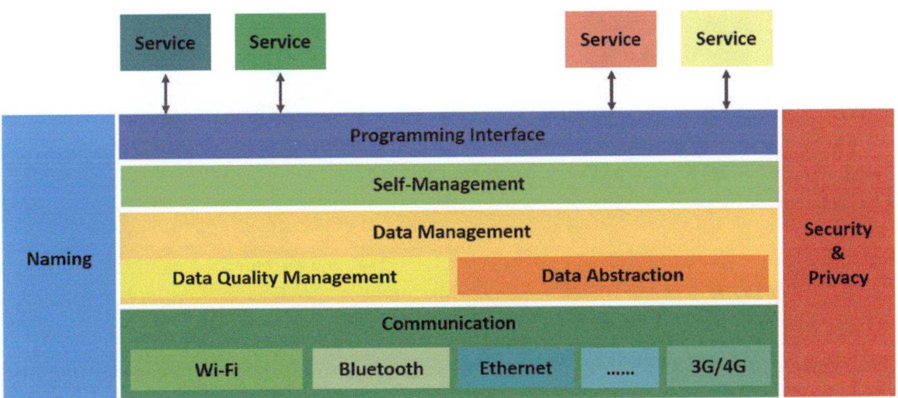

Fig. 2.3 The logical view of EdgeOS$_H$

In Fig. 2.3, we present the logical view of EdgeOS$_H$, which consists of four vertical layers: *Communication, Data Management, Self-Management,* and *Programming Interface*, as well as two extra components across all four layers, i.e., *Naming* and *Security & Privacy*. In the *Communication* layer, EdgeOS$_H$ needs to collect data from mobile devices and all kinds of things through multiple communication methods such as Wi-Fi, Bluetooth, ZigBee or a cellular network. Data from different sources need to be fused and massaged in the *Data Management* layer. In this layer, the data abstraction model will fuse and massage the data into one database, and the data quality model will manage the quality of the data. On top of the *Data Management* layer is the *Self-Management* layer. Services such as device registration, maintenance, and replacement will be provided here.

Moreover, *Self-Management* layer should also be able to detect the conflict among services and optimize the service quality. A unified programming interface should be supported to provide satisfactory performance for user applications with minimum development effort, which is the *Programming Interface* layer. The *Naming* mechanism is required for all layers with different requirements. Finally, data security and privacy should be protected in *Security & Privacy*.

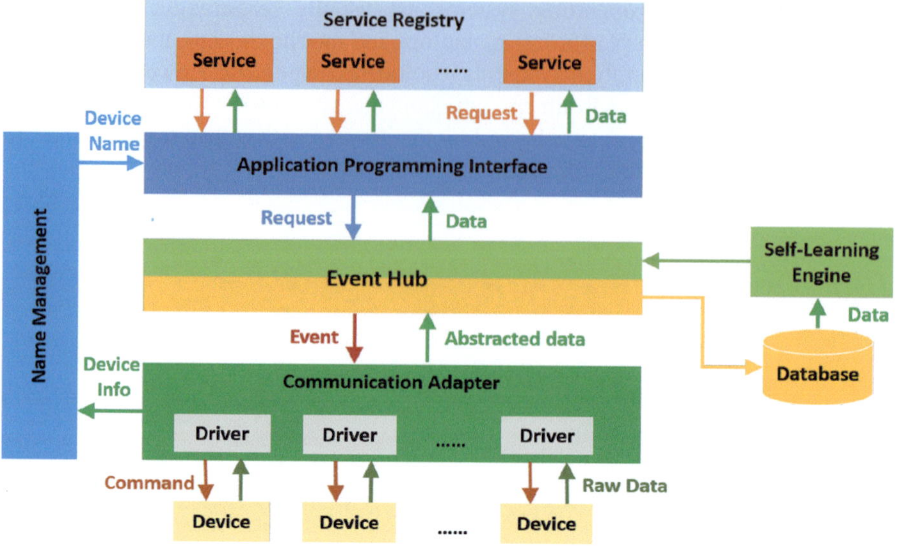

Fig. 2.4 The design of EdgeOS$_H$

2.3.2 Design

The design of EdgeOS$_H$ is shown in Fig. 2.4, including seven components: *Communication Adapter, Event Hub, Database, Self-Learning Engine, Application Programming Interface, Service Registry*, as well as *Name Management*, which stretches across other components. To integrate the device into EdgeOS$_H$, *Communication Adapter* gets access to devices by the embedded drivers. These drivers are responsible for sending commands to devices and collecting state data (raw data) from them. Sitting between devices and the *Event Hub, Communication Adapter* maps to the Communication layer in the logical view. It packages different communication methods that come from various kind of devices while providing a uniform interface for upper layers' invocation. In this way, developers and users do not need to deal with multiple kinds of communication methods when manipulating the system. Moreover, it only provides abstracted data to upper layer components, reducing privacy risk to some extent. As the core of the architecture, the *Event Hub* maps to two layers in the logical view: the *Data Management* and *Self-Management* layers. The *Event Hub* is responsible for capturing system events and sending instructions to lower levels. Those instructions are smart commands based on machine learning developed through communication with the Self-Learning Engine. It collects requests from services and sends them to the *Communication Adapter*, and in turn, collects abstracted data from the *Communication Adapter* and sends them to the upper layers. The *Database* is another component in the

Data Management layer. As a data-oriented system, EdgeOS$_H$ generates the large amount of data every day, which contains valuable information related to user preferences and settings. The *Event Hub* stores data in the *Database*. The data stored in the *Database* is utilized by the *Self-Learning Engine* that belongs to the *Self-Management* layer. The *Self-Learning Engine* creates a learning model. This learning model called the Self-Learning Model acts as an input to the *Event Hub* to provide decision-making capability. To provide better user experience, the *Self-Learning Engine* is developed to analyze user behavior, generate the personal model for the user, and help improve the system. *Application Programming Interface* (API) and *Service Registry* are located in the upper layers of the system and are utilized for third-party services. Developers are encouraged to use EdgeOS$_H$ APIs to communicate with the *Event Hub*, and register their services with the system. Required by all layers, *Name Management* helps the system keep devices organized. When a new device is registered with the system, *Name Management* allocates a name for it using the following rule: location (where), role (who), and data description (what). All layers compile this rule.

2.4 Summary

In this chapter, we presented our vision of a home operating system and introduced EdgeOS$_H$ to the domestic environment. We also listed several practical as well as non-functional challenges that should be addressed before our vision could be fulfilled.

We discussed the different components of EdgeOS$_H$ and introduced how EdgeOS$_H$ should be designed to have a stable programming interface and a self-management capability. We also discussed how the data should be abstracted, stored, and evaluated in EdgeOS$_H$. Security and privacy protection were also examined in this book as a home might be the most private environment for human beings. At last, we raised the lack of naming mechanism and discussed open issues such as a testbed for evaluating the performance of a smart home. We also discussed the user experience and the cost associated with a smart home.

We hope that EdgeOS$_H$ can be used as a guidance for prototyping on smart home systems. We also hope that this book can provide helpful information for researchers and practitioners from various disciplines when designing new technologies for smart homes.

In the next chapter, we will introduce the second case study as a framework for hybrid Cloud-Edge analytic. This framework fuses data from multiple stakeholders as the virtually shared dataset that is a collection of data and predefined functions by data owners. This framework will be breaking down into sub-services so that a user can directly subscribe to intermediate data and compose new applications by leveraging existing sub-services. An easy-to-use programming interface is also provided in the framework for both service providers and end users.

References

1. W. Shi, J. Cao, Q. Zhang, Y. Li, and L. Xu, "Edge computing: Vision and challenges," *IEEE Internet of Things Journal*, vol. 3, no. 5, pp. 637–646, 2016.
2. M. Satyanarayanan, "The emergence of edge computing," *Computer*, vol. 50, no. 1, pp. 30–39, 2017.
3. S. S. Intille, "Designing a home of the future," *IEEE pervasive computing*, vol. 1, no. 2, pp. 76–82, 2002.
4. J. A. Kientz, S. N. Patel, B. Jones, E. Price, E. D. Mynatt, and G. D. Abowd, "The georgia tech aware home," in *CHI'08 extended abstracts on Human factors in computing systems*.   ACM, 2008, pp. 3675–3680.
5. E. Soltanaghaei and K. Whitehouse, "Walksense: Classifying home occupancy states using walkway sensing," in *Proceedings of the 3rd ACM International Conference on Systems for Energy-Efficient Built Environments*.   ACM, 2016, pp. 167–176.
6. Y. Agarwal, B. Balaji, R. Gupta, J. Lyles, M. Wei, and T. Weng, "Occupancy-driven energy management for smart building automation," in *Proceedings of the 2nd ACM Workshop on Embedded Sensing Systems for Energy-Efficiency in Building*.   ACM, 2010, pp. 1–6.
7. D. Austin, Z. T. Beattie, T. Riley, A. M. Adami, C. C. Hagen, and T. L. Hayes, "Unobtrusive classification of sleep and wakefulness using load cells under the bed," in *Engineering in Medicine and Biology Society (EMBC), 2012 Annual International Conference of the IEEE*.   IEEE, 2012, pp. 5254–5257.
8. G. Gao and K. Whitehouse, "The self-programming thermostat: optimizing setback schedules based on home occupancy patterns," in *Proceedings of the First ACM Workshop on Embedded Sensing Systems for Energy-Efficiency in Buildings*.   ACM, 2009, pp. 67–72.
9. L. Shu, Y. Zhang, Z. Yu, L. T. Yang, M. Hauswirth, and N. Xiong, "Context-aware cross-layer optimized video streaming in wireless multimedia sensor networks," *The Journal of Supercomputing*, vol. 54, no. 1, pp. 94–121, 2010.
10. G. Zhang and M. Parashar, "Context-aware dynamic access control for pervasive applications," in *Proceedings of the Communication Networks and Distributed Systems Modeling and Simulation Conference*, 2004, pp. 21–30.
11. P. Liu, D. Willis, and S. Banerjee, "Paradrop: Enabling lightweight multi-tenancy at the network's extreme edge," in *Edge Computing (SEC), IEEE/ACM Symposium on*.   IEEE, 2016, pp. 1–13.
12. H. Zheng, H. Wang, and N. Black, "Human activity detection in smart home environment with self-adaptive neural networks," in *Networking, Sensing and Control, 2008. ICNSC 2008. IEEE International Conference on*.   IEEE, 2008, pp. 1505–1510.
13. J. Byun, B. Jeon, J. Noh, Y. Kim, and S. Park, "An intelligent self-adjusting sensor for smart home services based on zigbee communications," *IEEE Transactions on Consumer Electronics*, vol. 58, no. 3, 2012.
14. P. Rashidi and D. J. Cook, "Keeping the resident in the loop: Adapting the smart home to the user," *IEEE Transactions on systems, man, and cybernetics-part A: systems and humans*, vol. 39, no. 5, pp. 949–959, 2009.
15. C. Reinisch, M. J. Kofler, and W. Kastner, "Thinkhome: A smart home as digital ecosystem," in *Digital Ecosystems and Technologies (DEST), 2010 4th IEEE International Conference on*.   IEEE, 2010, pp. 256–261.
16. W. K. Edwards and R. E. Grinter, "At home with ubiquitous computing: Seven challenges," in *International Conference on Ubiquitous Computing*.   Springer, 2001, pp. 256–272.
17. S. Mennicken, J. Vermeulen, and E. M. Huang, "From today's augmented houses to tomorrow's smart homes: new directions for home automation research," in *Proceedings of the 2014 ACM International Joint Conference on Pervasive and Ubiquitous Computing*.   ACM, 2014, pp. 105–115.

Chapter 3
Firework: Data Analytics in Hybrid Cloud-Edge Environment

Now we are entering the era of the Internet of Everything (IoE) moreover, billions of sensors and actuators are connected to the network. As one of the most sophisticated IoE applications, real-time video analytics is promising to significantly improve public safety, business intelligence, and healthcare & life science, among others. However, cloud-centric video analytics requires that all video data must be pre-loaded To a centralized cluster or the cloud, which suffers from high response latency and a high cost of data transmission, given the scale of zettabytes of video data generated by IoE devices. Moreover, video data is rarely shared among multiple stakeholders due to various concerns, which restricts the practical deployment of video analytics that takes advantages of many data sources to make smart decisions. Furthermore, there is no efficient programming interface for developers and users to easily program and deploy IoE applications across geographically distributed computation resources. In this chapter, we present a new computing framework, Firework, which facilitates distributed data processing and sharing for IoE applications via a virtual shared data view and service composition. We designed an easy-to-use programming interface for Firework to allow developers to program on Firework. This chapter describes the system design, implementation, and programming interface of Firework. The experimental results of a video analytics application demonstrate that Firework reduces up to 19.52% of response latency and at least 72.77% of network bandwidth cost, compared to a cloud-centric solution.

3.1 Introduction

Cloud computing and edge computing are the core of future computing facilities and adopted in most data processing scenarios. However, an important or fundamental assumption behind them is that data is owned by a single stakeholder, where the

© The Author(s), under exclusive license to Springer Nature Switzerland AG 2018

J. Cao et al., *Edge Computing: A Primer*, SpringerBriefs in Computer Science,
https://doi.org/10.1007/978-3-030-02083-5_3

user or owner has full control privileges over the data. As we mentioned, cloud computing requires the data to be pre-loaded in data centers before a user runs its applications in the cloud [1], while edge computing processes data at the edge of the network but requires close control of the data producers and consumers. Data owned by multiple stakeholders is rarely shared due to various reasons, such as security concern (e.g., data across border), conflict of interest (e.g., data from competitors), privacy issue (e.g., data of health care), and resource limitation (e.g., extremely large and long network distance data transportation) and etc.

Taking the cooperation in connected health as an example, the health records of patients hosted by hospitals and customer records owned by insurance companies are highly private to the patients and customers and rarely shared. If an insurance company has access to its customers' health records, the insurance company could initiate personalized health insurance policies for its customers based on their health records. Another example is "find the lost" in city [2], where video streams from multiple data owners across the city are used to find a lost object. It is common that the police department manually collects video data from surveillance cameras on the streets, retailer shops, personal smartphones, or car video recorders in order to identify a specific lost object. If all these data could be shared seamlessly, it can save the tremendous amount of human work and identify an object in the real-time fashion. Furthermore, merely replicating data or running analyzing application provided by the third party on stakeholders' data may break the privacy and security restricts. Unfortunately, none of those mentioned above can be easily achieved by leveraging cloud computing or edge computing individually.

To attack barriers as mentioned earlier, *Firework* (**i**) fuses data from multiple stakeholders as the virtually shared dataset that is a collection of data and predefined functions by data owners. The data privacy protection cloud be carried out by privacy preserving functions preventing data leakage by sharing sensitive knowledge only to intended users; (**ii**) breaks down an application into subservices so that a user can directly subscribe to intermediate data and compose new applications by leveraging existing subservices; and (**iii**) provides an easy-to-use programming interface for both service providers and end users. By leveraging subservices deployed on both the cloud and the edge, *Firework* aims to reduce the response latency and network bandwidth cost for hybrid cloud-edge applications and enables data processing and sharing among multiple stakeholders. We implement a prototype of *Firework* and demonstrate the capabilities of reducing response latency and network bandwidth cost by using an edge video analytics application developed on top of *Firework*.

3.2 System Design

Firework is a framework for big data processing and sharing among multiple stakeholders in the hybrid cloud-edge environment. Considering the amount of data generated by edge devices, it is promising to process the data at the edge of the network to reduce response latency and network bandwidth cost. To simplify the development of collaborative cloud-edge applications, *Firework* provides a uniform programming interface to develop IoE applications. To deploy an application, *Firework* creates service stubs on available computing nodes based on a predefined deployment plan. To leverage existing services, a user implements a driver program and *Firework* automatically invokes the corresponding subservices via integrated service discovery. In this section, we will introduce the detailed design of *Firework*, including terminologies, system architecture, and programmability, to illustrate how *Firework* facilitates the data processing and sharing in the collaborative cloud-edge environment.

3.2.1 Terminologies

We first introduce the terminologies that describe abstraction concepts in *Firework*. Based on the existing definitions of the terminologies in our previous work [3], we extend and enrich their meanings and summarize them as follows:

- *Distributed Shared Data (DSD)*: Data generated by edge devices and historical data stored in the cloud can be part of the shared data. DSD provides a virtual view of the entire shared data. It is worth noting that stakeholders might have different views of DSD.
- *Firework.View*: Inspired by the success of object-oriented programming, a combination of *dataset* and *function*s is defined as a *Firework.View*. The *dataset* describes shared data and the *function*s define applicable operations upon the dataset. A *Firework.View* can be adopted by multiple data owners who implement the same functions on the same type of dataset. To protect the privacy of data owners, the *function*s can be carried out by privacy preserving functions that share sensitive data only to intended users [4].
- *Firework.Node*: A device that generates data or implements *Firework.View*s, is a *Firework.Node*. As data producers, such as sensors and mobile devices, *Firework.Node*s publish sensing data. As data consumers, *Firework.Node*s inherit and extend *Firework.View*s by adding functions to them, and the new *Firework.View*s could be further extended by other *Firework.Node*s.

  An example application could be the city-wide temperature data, in which scenario sensor data is owned by multiple stakeholders, and each of them provides public portals for data accessing. A user could reach all temperature data as if he/she operates on a single centralized data set. A sensor publishes a base *Firework.View* containing temperature data and reads function, and a *Firework.Node* can provide a new *Firework.View* that returns highest regional temperature by extending the base *Firework.View*.
- *Firework.Manager*: First, it provides centralized service management, where *Firework.View*s are registered. It also manages the deployed services built on top of these views. Second, it serves as the job tracker that dispatches

tasks to *Firework.Node*s and optimizes running services by dynamically scaling and balancing among *Firework.Node*s depending on their resource utilization. Third, it allocates computation resources including CPU, memory, network, and (optional) battery resources to running services. Fourth, it exposes available services to users so that they can leverage existing services to compose new applications.

- *Firework*: It is an operational instance of *Firework* paradigm. A *Firework* instance might include multiple *Firework.Node*s and *Firework.Manager*s, depending on the topology. Figure 3.1 shows an example of *Firework* instance consisting of five *Firework.Node*s employing heterogeneous computing platforms. If all *Firework.Node*s adopt homogeneous computing platform, such a *Firework* instance will be similar to cloud computing and edge computing.

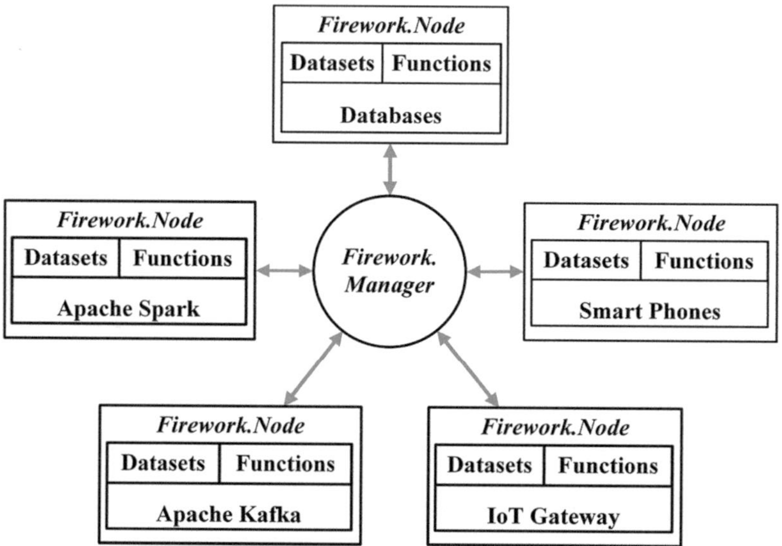

Fig. 3.1 An example of *Firework* instance that consists of heterogeneous computing platforms

3.2.2 Architecture

As a major concept of *Firework*, *Firework.View* is abstracted as a "class-like" object, which can be easily extended. *Firework.Node* can be implemented by numerous heterogeneous computing infrastructures, ranging from big data computation engines (e.g., Apache Spark [5], Hadoop [6], databases) and distributed message queues (e.g., Apache Kafka [7], MQTT [8], ZeroMQ [9], RabbitMQ [10]) in the cloud, to edge devices of smartphones and IoE gateways (e.g., Intel Edison, Raspberry Pi). *Firework.Manager* is the access point of a *Firework* instance that allows users to deploy and execute their services. Both *Firework.Node* and *Firework.Manager* can be deployed on the same computing node, where an edge node acts as not only a data consumer in the point of view of actuators but also a data producer in the point of view of the cloud.

To realize aforementioned abstractive concepts, we generalize them as a layered abstraction as shown in Fig. 3.2, which consists of *Service Management*, *Job Management*, and *Executor Management*. The *Service Management* layer performs service discovery and deployment, and the *Job Management* layer manages tasks running on a computing node. The combination of *Service Management* and *Job Management* fulfills the responsibilities of a *Firework.Manager*. The *Executor Management* layer, representing a *Firework.Node*, manages computing resources. In the following paragraphs, we will describe each layer in detail.

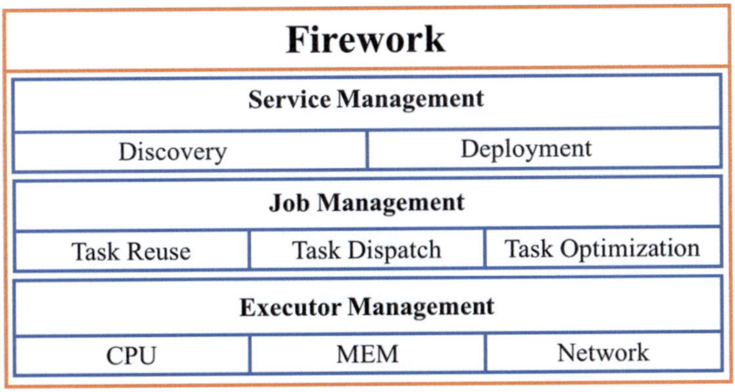

Firework		
Service Management		
Discovery	Deployment	
Job Management		
Task Reuse	Task Dispatch	Task Optimization
Executor Management		
CPU	MEM	Network

Fig. 3.2 An abstraction overview of *Firework*

Service Management To deploy a service on *Firework*, a user has to implement at least one *Firework.View* which defines the shared data and functions, and a deployment plan, which describes how computing nodes are connected and how services are assigned to the computing nodes. Note that the application defined deployment topology might be different from the underlying network topology. The reasons for providing customizable deployment plan are to avoid redundant data processing and facilitate application defined data aggregation. In a cloud-centric application, data is uploaded to the cloud based on a predefined topology, where a developer cannot customize the data collection and aggregation topology. However, in IoE applications, sensors/actuators (e.g., smartphones, on-vehicle cameras) change the network topology frequently, which requires the application deployment to be adapted depending on the available resource, network topology, and geographical location. Furthermore, *Firework.View* leverages multiple data sources to form a virtually shared dataset, where the data sources can be dynamically added or removed according to the deployment plan of the IoE application.

Upon a service (i.e., *Firework.View*) registration, *Firework.Manager* creates a service stub for that service (note that the same service registered by multiple nodes shares the same service stub entry), which contains the metadata to access the service, such as the network address, functions' entries, input parameters and etcetera. A service provider can create a *Firework.View* via extending a registered service and register the new *Firework.View* to another *Firework.Manager*. By chaining up all these *Firework.View*s, a complex application can be composed. Depending on the deployment plan, a service (i.e., *Firework.View*) can be registered to multiple *Firework.Manager*, and the same service can be deployed on more than one computing nodes. An application developer can implement services and similar deployment plans via the programming interfaces provided by *Firework*. We will show more details through a concrete example (i.e., video analytics implemented in *Firework*) in Sect. 3.2.3.

To take advantages of existing services, a user retrieves the list of available services by querying *Firework.Manager*. Then the user implements a driver program to invoke the service. Upon receiving the request, *Firework* filters out the computing nodes that implement the requested services. Afterward, *Firework* creates a new local job and dispatches the request to these computing nodes. Details about job creation and dispatch are explained in Sect. 3.2.2. By repeating this procedure, *Firework* instantiates a requested service by automatically creating a computation stream. The computation stream implements an application by leveraging computing resources along the data propagation path, which might include the edge devices and the cloud.

Considering the mobility and operational environment of edge devices, it is common that they may fail or change the network condition. To deal with failure or different network condition, *Firework* assigns a time-to-live interval to registered services and checks the liveness via heartbeat message periodically. A node will re-register its services after a failover or network condition change. When a node acting as *Firework.Manager* fails, it recovers all service stubs depending on persistent logs. Correctly, it rebuilds the connections based on the out-of-date service stubs and updates these service stubs if the connections are restored successfully. Otherwise, the service stubs are removed.

Job Management A user can send *Firework.Manager* a request to start a service. Upon receiving the invocation, a local job is created by the *Job Management* layer, which initializes the service locally. For each job, a dedicated communication port is assigned for exchanging control messages. Note that executors do not use this port for data communication. Next *Firework.Manager* forwards the request to available *Firework.Node*s that implement the *Firework.View* of the requested service. Lastly, the local job is added to the task queue waiting for execution. When the user terminates a job, the *Job Management* layer stops executors and releases the dedicated port of that job.

Firework provides elasticity of computing resource scaling via task reuse. In *Firework*, all services are public for all users, which potentially means that two different users could request the same service. In such situation, *Firework* reuses the same running task by dynamically adding an output stream to the task. It is worth noting that the input streams of service might come from different sources. Extra computing resources are allocated to a task if the resource utilization of an executor exceeds a threshold and vice versa. To reduce the I/O overhead of a task brought by communicating with multiple remote nodes, *Firework* uses a separate I/O manager, which will be introduced in Sect. 3.3, to perform data transmission so that a running service subscribes/publishes the input/output data from the I/O manager. In addition to the resource scaling, *Firework* also optimizes the workload among multiple nodes. A *Firework* node can inherit a base service without extending it. In this case, two consecutive nodes provide the same service. If the node closer to data sources is overloaded, it can delay and offload computation to the other node, which might be less loaded. The offload decision aims to minimize the response latency of the service, which depends on the resource utilization (e.g., CPU, memory, and network bandwidth).

Executor Management A task in *Firework* runs on an executor that has dedicated CPU, memory, and network resources. *Firework* nodes leverage heterogeneous computing platforms and consequently adopt different resource management approaches. Therefore, the *Executor Management* layer serves as an adapter that allocates computing resources to tasks. Specifically, some *Firework* nodes like smartphones or IoE gateways may adopt JVM or Docker [11], while some nodes like commodity servers may employ OpenStack [12] or VMWare, to host an executor. The *Job Management* layer fulfils the executor management and operated by the *Executor Management* layer.

3.2.3 Programmability

Firework provides an easy-to-use programming interface for both developers and users so that they can focus on programming the user-defined functions. An application on *Firework* includes two major parts: programs implementing *Firework.Views* and a driver program to deploy and interact with the application. A developer can decompose an application into subservices, such as data collecting on sensors, data preprocessing on edge, and data aggregation in the cloud. Each sub-service can be abstracted as a *Firework.View* and deployed on one or more computing nodes. By organizing them with a driver program, a user can achieve real-time analytics in the collaborative cloud-edge environment.

Specifically, *Firework* implements two basic programmable components, **FWView** and **FWDriver** that represent *Firework.View* and driver program, respectively. The *FWView* adopts a continuous execution model, in which a *FWView* continuously receives data from the input streams, processes the data, and sends out the data to other nodes. Listing 3.1 shows the Java code of *FWView*. As mentioned in Sect. 3.2.2, when a *Firework.View* is registered, a service stub is created, which contains the metadata information. Thus, the *FWView* could retrieve input and output streams from the *Job Management* layer, as shown by the *getFWInputStream()* and *getFWOutputStream()* functions in Listing 3.1. Note that,

```
/* FWView: an implementation of Firework.View */
public class FWView implements Runnable {
    protected String serviceName;
    protected FWInputStream[] inputStreams;
    protected FWOutputStream[] outputStreams;
    public FWView(String serviceName) {
        this.serviceName = serviceName;
        this.getFWInputStream();
        this.getFWOutputStream();
    }
    /* Get input streams from JobManager. */
    public void getFWInputStream() {
        inputStreams = JobManager.getInputStream(serviceName);
    }
    /* Get output streams from JobManager. */
    public void getFWOutputStream() {
        outputStreams = JobManager.getOutputStream();
    }
    /* Receive data from the input streams. */
    public Data[] read() {
        inputData = inputStreams.read();
    }
    /* Process the input data. */
    public Data[] compute(Data[] inputData) {
        // Do nothing by default.
        return inputData;
    }
    /* Send the processed data to other nodes. */
    public void write(Data[] outputData) {
        outputStream.write(outputData);
    }
    /* Run the computation procedure. */
    public void run() {
        while(true) {
            Data[] inputData = read();
            Data[] outputData = compute(inputData);
            write(outputData);
        }
    }
}
```

Listing 3.1 The Java code of FWView

the input streams of a service depend on the base services that provide input data for that service. The most important function of *FWView* is the *compute()*, in which a user implements the payload. By calling the *run()* function, a *Firework* node repeats the actions of data receiving (via *read()*), processing (via *compute()*), and sending (via *write()*). Since an application on *Firework* is decomposed into several subservices, a developer needs to implement multiple *FWView*s, which perform different functionalities.

The other basic programmable component of *Firework* is the *FWDriver*, which provides the capabilities to deploy and launch an application. In Listing 3.2, we illustrate the basic functionalities of an application driver. The *deploy()*, *start()*, and *stop()* functions allow users to manage their applications, and the *retrieveResult()* function pulls final outcomes from a *Firework.Manager*. All these functions are conducted by *FWContext*, which maintains a session between a user and a *Firework* instance. The *DeployPlan* is a supplemental component of *FWDriver*, which

```java
/* DeployPlan: the application defined topology. */
public class DeployPlan {
    private List<Rule> rules;
    /* Add a new rule to the deployment plan. */
    public void addRule(Rule rule) {
        rules.add(rule);
    }
}
/* FWDriver: Firework application driver. */
public class FWDriver {
    /* FWContext: creates a session between user and Firework.Manager */
    private FWContext fwContext;
    private DeployPlan deployPlan;
    private UUID uuid;
    private String serviceName;
    public FWDriver(String serviceName, DeployPlan deployPlan) {
        this.serviceName = serviceName;
        this.deployPlan = deployPlan;
        this.fwContext = new FWContext();
    }
    /* Deploy an application and get an UUID back. */
    public void deploy() {
        uuid = fwContext.configService(serviceName, deployPlan);
    }
    /* Launch a deployed service by uuid. */
    public void start(Parameter[] params) {
        fwContext.startService(uuid, params);
    }
    /* Stop an application by uuid. */
    public void stop() {
        fwContext.stopService(uuid);
    }
    /* Get the final results from Firework.Manager. */
    public Data[] retrieveResult() {
        return fwContext.retrieveResult(uuid);
    }
}
```

Listing 3.2 The Java code of FWDriver

describes an application-defined topology. Without providing a deployment plan, *Firework* uses the network topology as the default one, which might lead to redundant computation. A user can define a rule-based deployment plan to compose subservices as needed. An example of deployment plan could be grouping sensors by regional areas so that a single *Firework* node processes all data in the same region, which is straightforward for certain application scenarios, especially when the same stakeholder owns all sensors. However, when a user employs subservices owned by multiple stakeholders, the underlying network topology might not be able to aggregate all data to the user. Therefore, *Firework* provides the *DeployPlan* for users to customize the data propagation routes.

We use edge video analytics (i.e., search a targeted license plate) as an example to demonstrate how to program on *Firework*. In edge video analytics, we simplify the scenario and assume there are three nodes, including a camera, an edge node, and a cloud node. We also split the video search application into three subservices, of which the camera collects video data and sends it to the edge node; the edge node detects and recognizes a license plate from the video data and sends the

```java
/* VideoStream: implements data collection on a camera. */
public class VideoStream extends FWView {
    private SensorInputStream[] sensors;
    public VideoStream(String serviceName) {
        super(serviceName);
        this.getFWInputStream();
    }
    /* Get the input stream directly from sensors. */
    @Override
    public void getFWInputStream() {
        sensors = getSensorInputStream();
    }
    /* Read data from sensor instead of FWInputStream. */
    @Override
    public Data[] read() {
        Data[] inputData = new ArrayList<Data>();
        for (SensorInputStream aSensor : sensors) {
            inputData.add(aSensor.read());
        }
        return inputData;
    }
}
```

Listing 3.3 An example of VideoStream based on FWView

corresponding result to the cloud node if the targeted license plate is found; and the cloud node serves as a *Firework.Manager* and interacts with a remote user. We implement the entire application with three *FWView*-based services and a *FWDriver*-based client program. Listing 3.3 illustrates the video stream service on the camera, which performs data collecting. We rewrite the *getFWInputStream()* and *read()* functions since the data collection does not rely on other *Firework* service, in which case the camera directly collects data from onboard sensors and sends the data out without further computation (recall that by default *compute()* does not manipulate the data). On the edge node, we implement *VideoSearch* service as shown in Listing 3.4. We rewrite *compute()* function to perform the license plate detection and recognition. We implement *VideoAnalytics* service on the cloud node and omit the code since it uses default *FWView* implementation to forward the results to a user. As a user, a driver program is needed to invoke the service. Listing 3.5 shows an example of an application driver program that interacts with *VideoAnalytics* on the cloud node.

Through the above example of real-time video analytics (i.e., license plate recognition), we design a programming interface for *Firework* to make the framework easy to use. By separating the implementation of service and driver program, *Firework* allows a third party to leverage existing services by only providing a driver program. A user can also interact with an intermediate node (e.g., the edge node in above example) to leverage the semi-finished data to build his/her application.

```
/* VideoSearch: implements license plate recognition. */
public class VideoSearch extends FWView {
    public VideoSearch(String serviceName) {
        super(serviceName);
    }
/* Process data using user defined function. */
@Override
public Data[] compute(Data[] inputData) {
    Data[] outputData = LicensePlateDetectionAndRecognition(inputData);
    return outputData;
    }
}
```

Listing 3.4 An example of VideoSearch based on FWView

```
/* VideoAnalyticsDriver: a program to exploit the service on the cloud node. */
public class VideoAnalyticsDriver {
    public static void main(String[] args) {
        DeployPlan deployPlan = new DeployPlan();
        FWDriver fwDriver = new FWDriver("VideoAnalytics", deployPlan);
        fwDriver.deploy();
        Parameter[] params = Parameter.parse(args);
        fwDriver.start(params);
        Data[] results = fwDriver.retrieveResult();
        fwDriver.stop();
    }
}
```

Listing 3.5 An example of application driver

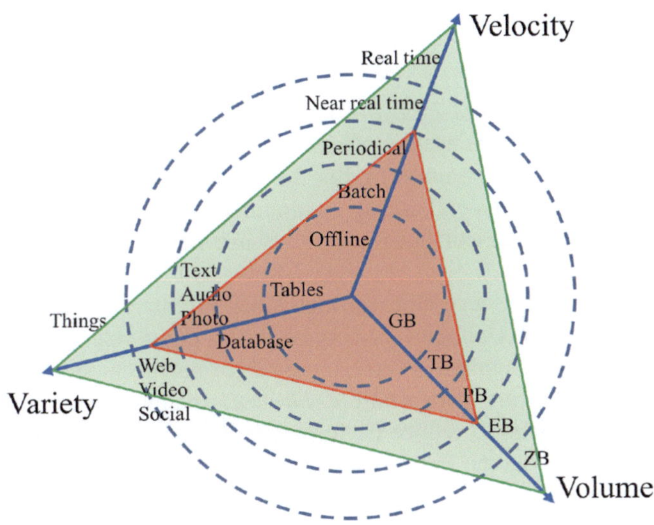

Fig. 3.3 The comparison of cloud computing and edge computing from the perspective of 3Vs. The red triangle represents cloud computing and the green triangle represents edge computing

3.2.4 Execution Model Comparison

We compare *Firework* with *Cloud Computing* and *Edge Computing* in terms of volume, variety, and velocity. As shown in Fig. 3.3, *Firework* extends the capabilities of cloud computing by leveraging edge computing, where the data volume is expanded due to the data generated by IoE devices and the low latency computation is achieved by pushing computation to data sources. Specifically, *Firework* distinguishes from cloud computing and edge computing in the following aspects: (i) *Firework* provides virtual data sharing among multiple stakeholders and data processing across the edge and the cloud. In contrast to *Firework*, cloud computing focuses on centralized computation resource sharing and data processing, and edge computing focuses manipulate local data with low latency and network bandwidth cost; (ii) *Firework* allows data owners to define the functions that can be performed on their data and shared with other stakeholders. The cloud computing collects data from users and defines the functions/services by the owners of clouds; (iii) *Firework* reduces the network bandwidth cost by performing the computation at data sources; and (iv) *Firework* leverages the cloud, as well as edge devices (and processing units placed close to the edge devices) so that the latency and network bandwidth cost can be reduced.

Up to this point, we have introduced the system design of *Firework* and illustrated the programmability of *Firework* via walking through the implementation of a potential IoE application on *Firework* and compared with existing computing paradigms. In the next section, we will explain the details of prototyping *Firework*.

3.3 Implementation

We implement a prototype of *Firework* using Java. Figure 3.4 shows an example architecture of our prototype system, which includes four *Firework* nodes and one *Firework* client. A *Firework* node fulfills the three-layered system design and a *Firework* client delegates an end user to communicate with *Firework* instance.

In the service management layer of a *Firework* node, the service registration is performed by the *Service Stub Manager* (shown in Fig. 3.4) built on *etcd* [13], which is a key/value store accessible through RESTful interface. When a service is registered on a *Firework.Manager*, the service access portal (e.g., service name and its IP address and port number) is stored in *etcd* for persistent storage, which will also be used for recovering from a failure. *Firework* maintains an in-memory copy of all the key/value pairs to reduce performance degradation caused by querying the *etcd* with REST requests. For the same service registered by multiple *Firework* nodes, we use the same *etcd* entry to store all the service access portals. To obtain the liveness of a registered service, *Firework* periodically sends heartbeat messages to all the portals and refreshes the time-to-live attributes and the list of live portals for the corresponding *etcd* entry. Another reason we choose *etcd* is that it provides a RESTful interface for a user to query available services. It is worth noting that users can query any *Firework.Manager* to retrieve available services and compose their applications. Another component in service management layer is the *Deployment Manager* (shown in Fig. 3.4), which decides if a *Firework* node satisfies the application defined deployment plan and informed job management layer to launch services.

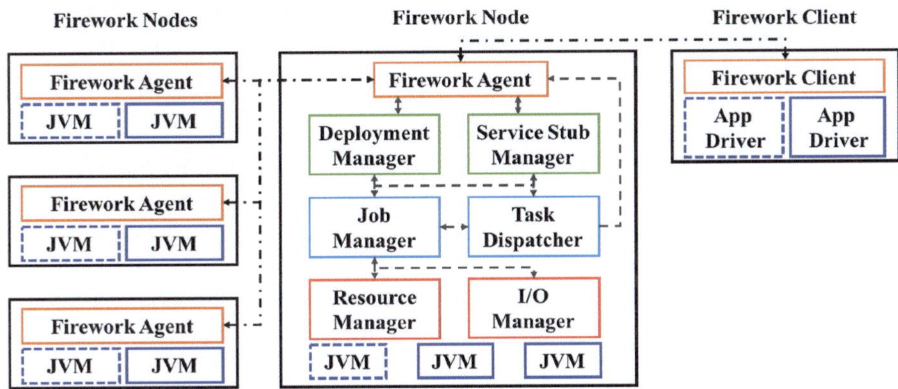

Fig. 3.4 An architecture overview of *Firework* prototype

In the middle layer of a *Firework* node, the *Job Manager* (shown in Fig. 3.4) is responsible for task decomposition, scheduling, and optimization. First, a job request is analyzed to determine its dependencies (i.e., the services relying on), and each dependency service is notified through the *Task Dispatcher* (optional) after applying the rules in the deployment plan. Then a local task is created and added to the task queue. In the current implementation, we use a first-come, first-serve queue for task scheduling. Finally, a task is submitted to an executor for execution. When multiple users request a service, *Firework* reuses current running task by adding output streams to the task, where a centralized I/O manager is used in the executor management layer (explained in next paragraph) for the stream reusing.

The bottom layer is the executor management layer, where we implement the *Resource Manager* and *I/O Manager* (shown in Fig. 3.4). When a task is scheduled to run, an executor (i.e., a JVM in our implementation) is allocated for it. It is worth noting that we can extend the *Resource Manager* to be compatible with other resource virtualization tools (e.g., Docker [11], OpenStack [12]) by adding corresponding adapter. The input and output of executors are carried out by the centralized *I/O Manager*, which is implemented as message queues. An executor subscribes to multiple queues as the input and output streams. The reasons we use a separated I/O manager are multifold. First, it is more efficient to manage the data transmission of an executor by dynamically adding or removing data streams to the message queue of the executor, which can be easily employed for task reuse. Second, by splitting the I/O management out from an executor, it reduces the programming efforts of developers so that they can focus on the functionalities. Third, such a design make it easy to leverage third-party message queuing systems (e.g., Apache Kafka [7] and MQTT [8]). When there is a huge number of sensors reporting to a single aggregation node, it makes *Firework* more scalable by simply adding more aggregation nodes and subscribing from the queuing systems that guarantee exactly-once data processing semantics. Fourth, a unified system level security protection can be applied on top of the I/O communication to guarantee data integrity and fidelity. Therefore, a separated I/O manager is used in *Firework*.

By deploying on multiple computing nodes, an instance of *Firework* system can be materialized. As shown in Fig. 3.4, multiple *Firework* nodes communicate with each other via the *Firework Agent* and form different topology based on an application-defined deployment plan. Note that we use star topology in Fig. 3.4 as an example topology of a *Firework* instance. A user can interact with *Firework* using the utilities provided by the *Firework Client* and deploy multiple applications (the solid-line rectangles and dashed-line rectangles in Fig. 3.4) on the same *Firework* instance.

Up to this point, we have introduced the implementation details of the prototype system of *Firework* (Fig. 3.5).

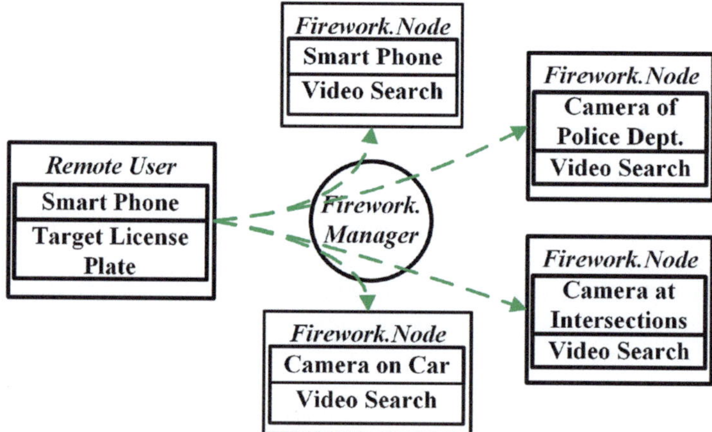

Fig. 3.5 A high level overview of a *Firework* instance for searching a target license plate in urban area

3.4 Discussion

In this chapter, we narrow down the scope of *Firework* to prototyping and programming interface implementation. In this section, we discuss potential issues and limitations of *Firework*, regarding system design and performance optimization.

Privacy Data captured by IoE devices can contain private information, e.g., GPS data, streams of video or audio, which might be used for complex analytics at somewhere other than where the data is generated. Thus, it is critical that, only data that is privacy complaint is sent to the edge or the cloud for further analysis. As we mentioned, *Firework* supports privacy preserving function, which can be adopted by implementing a function, such as a face blurring of video frames in [14, 15], as a predefined function of a *Firework.View*. Since privacy preserving function is attached to the shared subservices of each service owner, it is feasible for a downstream subservices to apply different privacy policies by extending existing subservices (i.e., extending a *Firework.View* to add/override existing privacy-preserving functions). Besides, *Firework* manages data communication using a separate I/O controller, where a natural security enhancement can be added by using secure communication protocols.

Fault Tolerant In the prototype of *Firework*, the *Job Manager* (shown in Fig. 3.4) tries to restart a job when the job fails due to software failure (e.g., out of memory, uncaught exceptions). However, the *Job Manager* cannot restart a job when the underly hardware fails. In our license plate recognition example, if all cameras fail and the VS is unavailable, the PD and PR are still running on the edge nodes and the cloud, but a user cannot get any output. In such case, *Firework* restores the VS whenever a camera is restored. Since *Firework* leverages computing resource that is owned by stakeholders and not controlled by *Firework*, there is no guarantee that an unavailable subservice would be available shortly. Thus, the fault tolerance in *Firework* depends on the underly fault tolerant mechanisms of stakeholders that might be very different. Thus, we leave the fault/failure detection in *Firework* as future work.

Optimization To simplify the scenario, we assume that an application can be decomposed and represented by a sequence of n functions, and m computing nodes are connected in a line. The goal of optimizing computation offload is to minimize the end-to-end latency by optimizing the allocation of n functions over m nodes, where the functions have to be allocated sequentially. In the current implementation of *Firework*, we use a simple deployment policy, in which the optimization target is a weighted sum of response latency and network bandwidth cost. Using the default deployment policy, it is possible to assign all subservices on one edge node (e.g., a smartphone), in which case the response latency (e.g., only including the time used for license plate detection and recognition) and network bandwidth cost(e.g., there is no data transmitted through network since all data are consumed locally on the smartphone) is minimized. However, it leads to high power consumption, which is infeasible and leads to short battery life. Furthermore, automatic functionality decomposition of an application increases the difficulty to optimize the function placement because the optimization goal of function decomposition might be contrasted to that of function placement. Therefore, we leave the automatic functionality decomposition and workload placement/migration as a future work so that *Firework* provides an efficient algorithm that co-optimizes these goals with little user intervention.

3.5 Summary

Real-time video analytics becomes more and more important to IoE applications due to the richness of video content and the huge potential of unanticipated value. To undertake the barriers of deploying IoE applications, we introduce a new computing framework called *Firework* that is a data processing and sharing platform for hybrid cloud-edge analytics. We illustrate the system design and implementation and demonstrate the programmability of *Firework* so that users can compose and deploy their IoE applications over various computing resources at the edge of the network and in the cloud. The evaluation of an edge video analytics application shows that *Firework* reduces response latencies and network bandwidth cost when using either LAN or LTE connection, compared to a cloud-centric solution. For the future work, we will explore automatic service/functionality decomposition so that *Firework* could dynamically optimize the subservice deployment according to the usage of computing, network, and storage resources on computing nodes.

Up to this point, we have presented the *Firework* framework that allows stakeholders and users to effectively and efficiently share and develop cloud-edge analytics applications. Combining with the H_2O, DyBBS, and *Firework*, we implement a series of systems that improve the performance of stream processing systems and facilitate the development of big data analytics in the cloud-edge environment.

In the next chapter, we will introduce a case study on video analytics on Edge computing platform.

References

1. M. Armbrust, A. Fox, R. Griffith, A. D. Joseph, R. Katz, A. Konwinski, G. Lee, D. Patterson, A. Rabkin, I. Stoica *et al.*, "A view of cloud computing," *Communications of the ACM*, vol. 53, no. 4, pp. 50–58, 2010.
2. W. Shi, J. Cao, Q. Zhang, Y. Li, and L. Xu, "Edge computing: Vision and challenges," *IEEE Internet of Things Journal*, vol. 3, no. 5, pp. 637–646, 2016.
3. Q. Zhang, X. Zhang, Q. Zhang, W. Shi, and H. Zhong, "Firework: big data sharing and processing in collaborative edge environment," in *Proceedings of the Workshop on Hot Topics in Web Systems and Technologies*, Oct 2016.
4. R. Agrawal and R. Srikant, "Privacy-preserving data mining," in *ACM SIGMOD Record*, vol. 29, no. 2.　ACM, 2000, pp. 439–450.
5. M. Zaharia, M. Chowdhury, M. J. Franklin, S. Shenker, and I. Stoica, "Spark: cluster computing with working sets," in *Proceedings of the 2nd USENIX Conference on Hot Topics in Cloud Computing*, vol. 10, 2010, p. 10.
6. K. Shvachko, H. Kuang, S. Radia, and R. Chansler, "The hadoop distributed file system," in *Proceedings of IEEE 26th Symposium on Mass Storage Systems and Technologies (MSST)*. IEEE, 2010, pp. 1–10.
7. J. Kreps, N. Narkhede, and J. Rao, "Kafka: A distributed messaging system for log processing," in *Proceedings of the NetDB*, 2011, pp. 1–7.
8. U. Hunkeler, H. L. Truong, and A. Stanford-Clark, "Mqtt-s: A publish/subscribe protocol for wireless sensor networks," in *Communication Systems Software and Middleware and*

Workshops, 2008. COMSWARE 2008. 3rd International Conference on, Jan 2008, pp. 791–798.

9. P. Hintjens, *ZeroMQ: Messaging for Many Applications.* "O'Reilly Media, Inc.", 2013.
10. "RabbitMQ," https://www.rabbitmq.com/, [Online; accessed Dec. 1st, 2016].
11. (2017, Mar.) Docker. [Online]. Available: https://www.docker.com/
12. "Openstack," https://www.openstack.org/edge-computing/.
13. (2016, Sep.) etcd. [Online]. Available: https://github.com/coreos/etcd
14. P. Simoens, Y. Xiao, P. Pillai, Z. Chen, K. Ha, and M. Satyanarayanan, "Scalable crowd-sourcing of video from mobile devices," in *Proceedings of the International Conference on Mobile systems, Applications, and Services.* ACM, 2013, pp. 139–152.
15. J. Wang, B. Amos, A. Das, P. Pillai, N. Sadeh, and M. Satyanarayanan, "A scalable and privacy-aware iot service for live video analytics," in *Proceedings of the Multimedia Systems Conference.* ACM, 2017, pp. 38–49.

Chapter 4
Distributed Collaborative Execution on the Edges and Its Application on AMBER Alert

As the entering of the era of the Internet of Everything (IoE), billions of geographically distributed things will connect to the Internet, and the things will generate hundreds of Zettabytes data per year. Pushing all of those data to the cloud leads to tremendous network bandwidth cost and latency, which is hard to be accepted for some latency-sensitive applications, such as vehicle tracking using city-wide cameras, which is promising to improve the AMBER alert system. Edge computing as an emerging computing paradigm can significantly reduce data transmission and response latency for latency-sensitive applications by processing data at the proximity of data sources.

However, most vision-based analytics is computationally intensive, and an edge device might be overwhelmed given tens of frames each second for real-time analyzing. Furthermore, a customized and flexible interface is required to implement efficient tracking strategies.

In this chapter, we extend a big data processing framework *Firework*, to support the collaboration between multiple edge devices and customizable task scheduling strategy. Based on the extended *Firework*, we implemented the AMBER Alert Assistant (A3 in short), which can efficiently track and locate a suspect vehicle by analyzing the city cameras' data in the real-time fashion. We also propose two kinds of customized task scheduling algorithms for vehicle tracking in A3. The comprehensive evaluation results show that A3 can achieve real-time video analytics by collaboration among multiple edge devices and the proposed location-direction-related diffusion strategy is very useful in controlling the searching area for vehicle tracking by the smart selection of candidate cameras.

J. Cao et al., *Edge Computing: A Primer*, SpringerBriefs in Computer Science,
https://doi.org/10.1007/978-3-030-02083-5_4

4.1 Introduction

The last decade, researchers and practitioners have treated cloud computing [1] as the de facto large-scale data processing platform. Numerous cloud-centric data processing platforms [2–7] that leverage the MapReduce [8] programming framework, have been proposed for both batched and streaming data. In recent years, we have witnessed the onset of the Internet of Everything (IoE) era [9], where billions of geographically distributed sensors and actuators are connected and immersed in our daily life. As one of the most sophisticated IoE application, real-time video analytics promises to improve public safety significantly. *Video Analytics* leverage information and knowledge from video data content to address a particular applied information processing need; and as the public safety community massively adopts this technology, it provides near real-time situational awareness of citizens' safety and urban environments, including automating the laborious tasks of monitoring live video streams, streamlining video communications and storage, providing timely alerts, and making the task of searching enormous video archives tractable [10]. However, the conventional cloud-centric data processing model is inefficient to process all IoE data in data centers especially for the response latency, network bandwidth cost, and possible privacy concerns, given the zettabytes of data generated by edge devices [11, 12]. On the other hand, rarely are data shared among multiple stakeholders—a variety of concerns restrict video analytics' practical deployment, even though video data analytics could take advantage of many data sources to make a smart decision. Moreover, there is no efficient data processing framework for the community to program and deploy easily for public safety applications across geographically distributed data sources.

Beyond processing data in real time, this distributed and collaborative application in an edge-cloud environment also offers increased reliability, leading to a quicker response. Take, for example, America's Missing Broadcast Emergency Response (AMBER) alert system. Right now, tracking a suspect's vehicle heavily relies on the reports of witnesses. However, what if we leveraged edge computing instead, so a license-plate-recognition (LPR) application running city-wide security cameras could significantly improve the efficiency of suspect vehicle tracking? Then, the local license-plate-recognition process for an image could respond in milliseconds without data transmission, instead of waiting for seconds to communicate with a remote data center, which is far quicker than the time required to send the image to the data center, process the data, and retrieve the results. The potential consequence is stark; this could mean the difference between identifying the missing child immediately versus losing the child in sight. Moreover, the network traffic caused by sending a deluge of data to a data center significantly impacts network performance, which further intensifies the response latency and data transmission cost.

To realize the vision of edge computing and real-time video analytics for public safety, we must tackle several barriers systematically: First, no existing programming tool and framework allows programmers to build cost-effective real-time applications among various geographically distributed data sources. Second, most video analytics algorithms undoubtedly are computationally intensive, of which hundreds of milliseconds might be taken to process one video frame on edge nodes [13], so that the edge device is overwhelmed given the tens of frames in a second. Thus, the collaboration among multiple edge nodes would ensure high resource usage and provide opportunities to balance the workload. However, realizing efficient resource management and task scheduling are difficult in an edge computing environment, where the edge nodes primarily are heterogeneous, with various types of network connectivity. Third, efficient task scheduling might require user intervention, in which domain knowledge could be applied to boost the performance. However, rarely is user intervention considered in task allocation, mainly when the application-defined network topology differs from the physical network topology.

To tackle the issues mentioned above, we extend our previous work on *Firework* [14, 15]. Correctly, we implement a collaborative mechanism among multiple edge nodes for real-time data processing and a programming interface to construct a customized task-scheduling strategy, depending on the application-defined network topology. Based on *Firework*'s extensions, we implement the *AMBER Alert Assistant* (A3) system, which aims to improve target tracking efficiency (that is, tracking suspect's vehicles) using the *AMBER Alert* system. By implementing multiple customized task-scheduling strategies, we evaluate various target-tracking strategies relying on the status of real-time video analysis results.

The rest of the chapter is organized as follows. We describe our motivation in Sect. 4.2 and present the A3 system in Sect. 4.3. We evaluate the performance of A3 system in Sect. 4.4. Section 4.4.1 focuses on experiments and results. We review related work in Sect. 4.5. Finally, we conclude in Sect. 4.6.

4.2 Motivation

4.2.1 AMBER Alert

The *AMBER Alert* is a system that alters the public to child abductions that are implemented with different names in different countries [16]. In the United States, when a kidnapping occurs, an alert is sent to citizens' smartphone immediately if they are near the event location. The alert usually includes descriptive information about this event, such as time, location, the kidnapper's vehicle license plate number, and a description of the child or kidnapper. Then, witnesses can provide pertinent information to the police department if they spot the kidnapper's vehicle on the road. However, tracking a suspect's vehicle heavily relies on the reports of witnesses, which is inefficient, because many people miss the alert or might not recognize the suspect's vehicle.

Nowadays, video surveillance is ordinary in cities, including security cameras, traffic cameras, and even smartphone/on-dash cameras. For example, participants in

Project Green Light [17] provide their video data for public safety, which includes more than a hundred cameras in Detroit, Michigan. Using an automatic license-plate-recognition (ALPR) technique, video surveillance dramatically improves the tracking of kidnappers' vehicles. Coped with the privacy issues, other cameras (i.e., the cameras in the Project Green Light, on-dash car cameras, smartphone cameras) might provide video data by the owners.

Usually, the network bandwidth requirements of a 720P, 1080P, and 4K live video are 3.8 Mbps, 5.8 Mbps, and 19 Mbps, respectively. Thus, pushing all video data to the cloud leads to huge data transmission costs and high latency. Considering the amount of data generated by many cameras simultaneously, cloud-based solutions are no longer suitable for real-time video analytics, due to high data transmission costs, bandwidth requirements, and excessive latency. Edge computing processes data locally, which significantly reduces the data transmission cost and lowers network bandwidth requirements. However, it still has two barriers that prevent real-time vehicle tracking in the *AMBER Alert* system.

Limitations of Edge Devices

Most computer vision algorithms are computationally intensive, such as object detection, face recognition, and optical character recognition (OCR). Thus, the edge node might lack the computation resources needed to process video in real time. For vehicle tracking, an open source ALPR system is built, called OpenALPR [18], which usually has two stages: license plate detection and license recognition. The latter usually employs the OCR technique. Because of its video streaming features, it is worthwhile to consider video decoding; we also implemented motion detection using OpenCV [19], which detects the different areas between two frames and avoids needless license plate detection and recognition. Here, we measure OpenALPR's [18] performance on different devices, including the Amazon Elastic Compute Cloud t2 node (Amazon EC2 t2: an Amazon virtual machine with Intel Xeon CPU at 2.4 GHz), Dell Inspiron 5559 (a laptop with Intel i7-6500U at 2.5 GHz, going up to 3.1 GHz), Dell Wyse (a home gateway with Intel N2807 at 1.58 GHz) and Dell OptiPlex (an Intel i5-4590 at 3.3 GHz, going up to 3.7 GHz).

Figure 4.1 shows that an ALPR system without motion detection will cost much more than the one with motion detection executing on an Amazon EC2 t2 node. This is because motion detection marks the moving area between two frames, which significantly reduces the workload of the latter steps by reducing the recognition area. The average LPR time without motion detection is 191.05 ms, and the time with motion detection is 131.18 ms on an Amazon EC2 t2 node. The average ALPR's complete processing times for two cases are 195.69 ms and 187.59 ms. The difference between the two cases is small, which is why we use video from a peak period. We also use another set, where the time is less than 100 ms for instance with motion detection. In this chapter, we always use peak-period video data, because it is essential to consider the worst case to avoid overload. Thus, video analytics must be

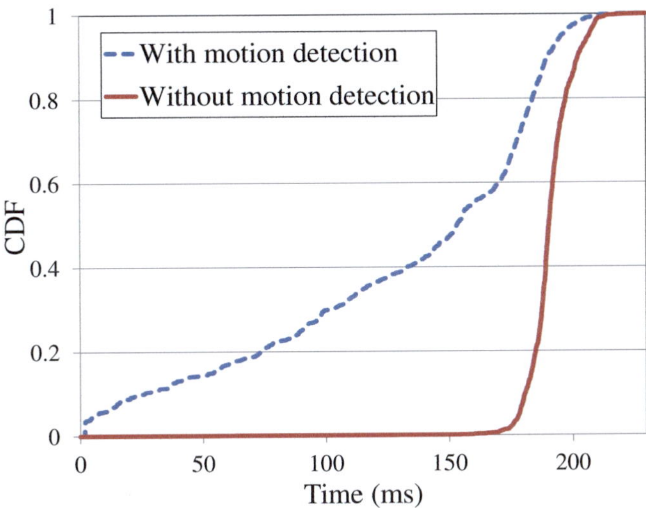

Fig. 4.1 The time cumulative distributive function (CDF) of license plate detection and recognition on an Amazon EC2 t2 node

multithreaded and have motion detection; both are included in our system design. Also, because LPR takes much less time to process than license plate detection, here we merely use LPR rather than the full steps of license plate detection and recognition.

Moreover, plate detection contributes the majority of processing time. All of the experimental devices can not analyze the video in real time by themselves if only having one worker thread. For guiding our design, we also get the cumulative distribution function (CDF) figure about this processing on the Amazon EC2 t2 node as shown in Fig. 4.1. As mentioned before, the motion detection has marked the motion area in frames and thus, the plate detection, and recognition only needs to process the marked area which saves a lot of processing time. The result of CDF also verifies this point. From the figure, we can know, the most significant processing is more than two hundred milliseconds, and there exists tens of percent of frames that do not need a plate detection and recognition, or part of an area in one frame need to be detected. It means the execution of plate detection and recognition is related to the particular scenario after using motion detection.

Control of the Vehicle Tracking Area

As the suspect's vehicle moves, the vehicle tracking area should be enlarged; and it also should shrink (or close in on) the area once the vehicle is found. A simple method to track a vehicle is to enlarge the radius of the area as time goes by. However, the speed of different routes varies. For example, the highway allows twice the speed of city streets. Thus, it is unreasonable to set the same rate for different types of cameras. Furthermore, mobile cameras deployed on taxis and garnered by crowdsourcing private vehicles could be integrated into the system to provide more video sources. It is exceedingly difficult, though, to provide a customized and dynamic control strategy for tracking the suspect's vehicle using both static and dynamic data provided by moving cameras.

4.2.2 Distributed Collaborative Execution on the Edge

As mentioned previously, there are two barriers to combining edge computing with the *AMBER Alert* system: the limitations of edge devices, and control of the vehicle tracking area. These can be abstracted to two types of collaborations—the collaboration of data processing, and the collaboration of task diffusion. This leads to the requirements for a distributed collaborative execution. Similar requirements exist in other edge computing applications. For example, consider activity detection in a jail. By analyzing captured videos, the threatening event can be detected using several features—such as the same people crowding in several different videos. This scenario involves synthetically analyzing multiple video data at different times to find a suspected event, which also needs to combine multiple edge nodes and the task control on multiple edge nodes. However, no platform or framework currently exists that can cope with both barriers. The *Firework* framework might be approximated for this case, and it provides a uniform programming interface to develop applications in the collaborative edge-cloud environment.

4.3 AMBER Alert Assistant

After extending the previous version of *Firework*, we implement an application called *AMBER Alert Assistant* (A3). After receiving an AMBER alert from police, A3 automatically tracks the suspect's vehicle by analyzing the video data of city-wide cameras and it also automatically controls the tracking area. In this section, first, we consider A3's potential use and application design. Then, we implement the application and discuss two task-scheduling strategies, which are used to control the vehicle tracking area. Note that here, a *Task* refers to a vehicle-tracking *Job*, and when a *Task Receiver* receives a *Task*, it creates a *Job* and launches related subservices.

4.3.1 Application Scenario

In A3, we consider an edge computing network model (see Fig. 4.2) inspired by Project Green Light [17], in which cameras are located at gas stations or shops and most of them connect to several edge devices, such as desktops. We also consider mobile cameras in A3, such as car's dash cameras, where the number of mobile cameras will increase quickly with the rise of autonomous driving. Each road camera and its edge nodes connect to a router via wired or wireless links, where the router is used to connect with a vast area network (WAN), and mobile cameras take part in A3 by connecting to a wireless cellular network. Once the edge node receives an alert from police, it will automatically pull the video from a connected camera and then collaboratively analyze video data with other local edge nodes in real time. As time passes, the edge node will extend the searching areas using a task-scheduling strategy. When the targeted vehicle is found, the edge node will automatically shrink the searching area to minimize resources, including energy.

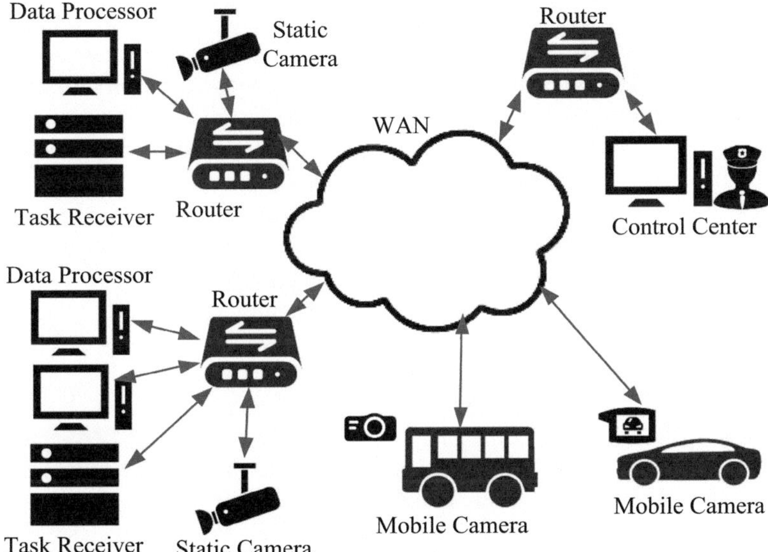

Fig. 4.2 The network settings of an A3 application scenario

4.3.2 Application Design

Based on A3's network, we define three types of devices following their functions in A3: the *Control Center*, *Task Receiver*, and *Data Processor*. These three types of devices are implemented based on our extended version of *Firework*, and we implement different *Firework.View*s for different functions.

Control Center

A *Control Center* is used by police to publish AMBER alerts to *Task Receiver*s and collect the reports from *Task Receiver*s. It configures a customized task scheduling by defining an application-defined topology. After starting a vehicle tracking task, it will receive regular status reports from a working node. Thus, it knows the current searching area, updated periodically. Once the targeted vehicle is found, it will receive the report from the related *Task Receiver*. After arresting the kidnapper, it clears the alert for all *Task Receiver*s.

Task Receiver

The video data captured by a traffic camera will be pulled and analyzed by a local *Task Receiver*. As an initial operation, it will get the application-defined topology from the *Control Center*, and store it in the *Service Deployment* module using JSON format. As Listing 4.1 shows, for a video searching service, it has a field called *timeout*, which it uses to control the opportunity for transferring the task to an appointed edge node. It also defines a topology for the *Data Processor*, to collaboratively analyze video in real time. According to the values watched by the *Task Monitor* (e.g., the message queue's length), a *Task Receiver* dispatches part of the video analytics workloads (such as plate recognition) to the *Data Processor*s defined in the application-defined topology. Once the *Task Receiver* get a report from others or a stop signal from the *Control Center*, it will stop analyzing the video and clear the message queue waiting for processing. Individually, it will forward the report to the *Task Receiver* that transferred this task initially.

Data Processor

The *Data Processor* in A3 only provides services to analyze the video data, but it is necessary for real-time video analytics. This type of edge node includes any *Firework.Node*-hosted local device, such as a desktop, laptop, and even smartphone. After accessing the same network with the *Task Receiver*, it gets the sub video analytics task from the connected *Task Receiver*.

Note that the connections between edge nodes are defined in the *Service Deployment* module, and controlled by the *Access Control* module. This means that the connection from an unknown edge node (that is, a node not defined in an application-defined topology) will be rejected. In Fig. 4.2, the police's desktop is a *Control Center* node, and other devices are either a *Task Receiver* or a *Data Processor*, where only one *Task Receiver* exists.

4.3.3   Implementation Details

As we mentioned, we implemented three types of *Firework.Node*s: *Control Center*, *Task Receiver*, and *Data Processor*. The *Task Receiver* and *Data Processor* provide (sub)services about video analytics. In this section, we will use the Java interface to describe A3's different services. Note that we use the word "interface" instead of "(sub)service".

Listing 4.1 Example of the *Control Center*'s interface

```
/* ControlCenter: the interface should be implemented on the edge node located
    at the police department */
public interface ControlCenter_I{
    /* Send Video Searching Task to edge nodes */
    public void PublishAMBERAlert(List<int> node_list, JsonString
        vehicle_info);
    /* Stop Video Searching Task */
    public void ClearAMBERAlert(int task_id);
    /* Report object location */
    public void TaskReport(String task_report);
    /* Set the application defined topology */
    public void SetTopology(JsonString whole_topology);
    /* Get the application defined topology */
    public JsonString GetTopology(int node_id);
    /* Report status from Task Receiver */
    public void ReportStatus(JsonString status);
    /* Login for mobile cameras */
    public Byte[] MobileCameraLogin(Byte[] data);
    /* Logout for mobile cameras */
    public Byte[] MobileCameraLogout(Byte[] data);
}
```

Listing 4.1 illustrates the services we implemented in this type of *Firework.Node*.
The `PublishAMBERAlert` interface and `ClearAMBERAlert` interface allow
the police to publish or clear an alert to the related *Task Receiver*. Once an edge node
finds the targeted vehicle, the edge node will report the vehicle's location using
the `TaskReport` interface. As we mentioned in Sect. 4.2, there are hundreds of
cameras in the city, so we need a dynamic topology to optimize the task-transfer
scheme. We define and implement the `SetTopology` and `GetTopology` interfaces to control the application-defined topology for *Task Receiver*s. We provide
the `MobileCameraLogin` and `MobileCameraLogout` interfaces for mobile
camera login and log out.

Listing 4.2 shows a *Task Receiver* provides several services. The *Control
Center* will call the `FindObject` interface to publish an AMBER alert. When
a *Task Receiver* receives a task from the *Control Center*, it will set up a timer
for task scheduling according to the value of timeout in the service deployment.
Moreover, once conditions are met, a task will be transferred to other *Task Receiver*s
by calling the `TaskTransfer` interface. If a *Task Receiver* receives a stop
signal from the `StopSearching` interface, it will stop searching for this alert
task, and also, it will forward this signal to the edge nodes that sent/received a
transferred task to/from the former. Once the targeted vehicle is found, the *Task
Receiver* will report the vehicle's location using the *Control Center*'s `TaskReport`
interface and it will send a clear sign to neighboring *Task Receiver*s. Then, it will
reset all task-scheduling timers for task scheduling. The `MotionDetection`,
`PlateRecognition` and `PlateDetection` interfaces are used to implement
video analytics. We implemented them based on the OpenALPR [18] Library
using message queues as the input and output streams. This means that the
`MotionDetection` interface will get the video frame from its input message

Listing 4.2 Example of a *Task Receiver*'s interface

```
/* TaskReceiver: the interface should be implemented on the edge node
     connected with control center */
public interface TaskReceiver_I{
    /* Receive object information for finding */
    public void FindObject(int task_id ,String object_info);
    /* Stop to search the object */
    public void StopSearching(int task_id);
    /* Receive object information for finding transferred by the other edge
       node */
    public void TaskTransfer(String object_info);
    /* Motion detection for a video */
    public Byte[] MotionDetection(Byte[] videoframe);
    /* Recongnize the license plate number from a license plate image */
    public Byte[] PlateRecognition(Byte[] image);
    /* Detect the license plate from motion of a frame */
    public Byte[] PlateDetection(Byte[] data);
}
```

Listing 4.3 Example of a *Data Processor*'s interface

```
/* DataProcessing: the interface should be implemented on the data processing
     edge node */
public interface DataProcessing_I{
    /* Recongnize the license plate number from a license plate image */
    Byte[] PlateRecognition(Byte[] data);
    /* Detect the license plate from the motion of a frame */
    Byte[] PlateDetection(Byte[] data);
}
```

queue, the data of which comes from a camera's live video. After processing, the `MotionDetection` interface will save the motion area image to its output message queue, which is also the input of the `PlateDetection` interface. As we mentioned, a subservice launches several worker instances for parallel execution. The `PlateDetection` interface supports this feature by declaring the deployment plan in a JSON string. The reason the `PlateRecognition` interface is not auto-scaling is because plate recognition is not as computationally intensive as plate detection, and some frames may not have a license plate.

The edge node of *Task Receiver* may does not have enough computing power to process frames in real time. Therefore, we set up several *Data Processors* in the local edge environment. Listing 4.3 shows the interface of a *Data Processor*. It just provides `PlateRecognition` and `PlateDetection` interfaces.

4.3.4 Task Scheduling

Here, we introduce the strategies used in the *Job Schedule* module to control the vehicle-tracking area in our implementation. Cameras have many limitations (for example, focal length, focus, and angle), and in the actual scenario, it usually

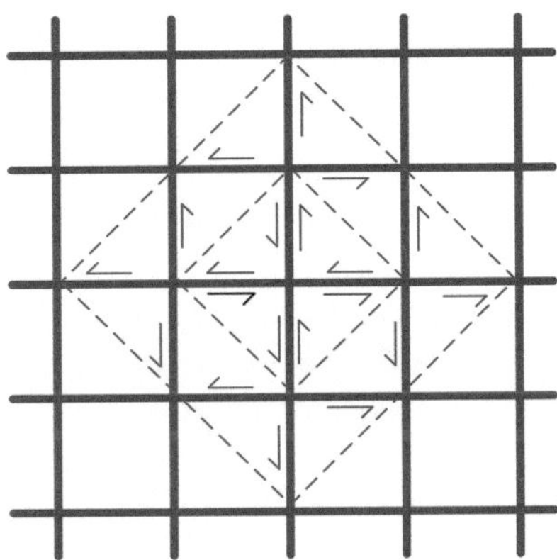

Fig. 4.3 A sample topology of road cameras

captures only one-way traffic flow. Thus, a two-way road needs two cameras located on each side of the road. In this case, we assume a simple and regular road topology where all roads are two-way and monitored by cameras (see Fig. 4.3). For a vehicle-tracking task, we present two diffusion models for task scheduling: distance-related diffusion (DD) and location-direction-related diffusion (LD). With both, we assume that a task will be sent to the edge node nearest to the kidnapping location with a fleeing direction. Thus, the number of initialization camera is $n_0 = 1$. For convenience, we choose the first crossing of the fleeing direction as the initial point for diffusing. Once the targeted vehicle is found, it will be reset so that only one edge node receives this task. For simplicity, we consider all cameras to have the same value of timeout for diffusing a task. The black arrow indicates the kidnapping location and driving direction.

Distance-Related Diffusion

ADD strategy, as the most straightforward strategy, diffuses the task according to the distance from where the kidnapping occurred. For example, the tracking area's radius will increase by a fixed number as time passes. For the topology in Fig. 4.3, the *Task Receiver* will send the task to seven other neighboring cameras based on its direction when diffusing a task. As Fig. 4.3 shows, the camera at the black arrow will transfer the task to the cameras in the area enclosed by the blue dashed square. Then, in the next cycle, all cameras in the green square will execute the task. According to the rule of diffusion, we see that the increased cameras are all located between two squares as the blue and green squares. Hence, the increment of two next times can be expressed by the equation $\Delta_t^D = sr(2t - 1), t \geq 2$, where s defined as the side numbers of the square is 4, and r defined as the cameras of each side is 2.

Thus, the number of working *Task Receiver*s for time t is expressed by the following equation:

$$n_t^D = \begin{cases} n_0, & t = 0 \\ srt^2, & t > 0 \end{cases} \tag{4.1}$$

Location-Direction-Related Diffusion

In the actual scenario, the DD strategy has many disadvantages. For instance, the speed on different roads varies (the speed on the highway is twice the speed allowed on city streets). So, if we set the tracking area's radius according to the highway speed, we waste the computing of local street cameras. On the other hand, if set the radius according to the local, street cameras, our tracking could fail once the kidnapper flees by the highway. We, therefore, propose an LD strategy for A3. In this model, the edge node will transfer the task to edge nodes according to the road topology. For example, the edge node located at the black arrow will transfer the task to the blue ones, because a vehicle in a crossroad has only four choices: go straight, or turn left, right, or around. As with the DD, the initial number of working edge nodes is $n_0^L = n_0$, and in the next cycle, it will include the black arrow and blue arrows shown in Fig. 4.3. In the second cycle, it will include the black arrow, blue arrows, and green arrows. Hence, the increment of two adjacent times can be expressed by the equation $\Delta_t^L = s\frac{r}{2}(2t - 1) + s\frac{r}{2}(2(t - 1) - 1), t \geq 3$.

Thus, the number of working *Task Receivers* each time t is expressed by the following equation:

$$n_t^L = \begin{cases} n_0, & t = 0 \\ n_0 + s\frac{r}{2}, & t = 1 \\ srt^2 - srt + s\frac{r}{2}, & t \geq 2 \end{cases} \tag{4.2}$$

Figure 4.4 shows the number of working nodes for the two strategies as mentioned earlier, where the time cycle is based on Eqs. (4.1) and (4.2). The results in Fig. 4.4 show that the LD launches fewer edge nodes to track the targeted vehicle, which will save significant computing resources and energy. The cameras in the city will not be shown regularly in Fig. 4.3, and neither will the road topology. Both strategies are easy to implement in *Firework*, to modify the application-defined topology. Besides, it is easy to set different timeout values for different edge nodes for optimization. For example, setting the timeout value to 30 seconds for highway cameras and 60 seconds for local cameras is more reasonable and efficient. Thus, the LD strategy will is more efficient because it significantly reduces the number of participant edge devices.

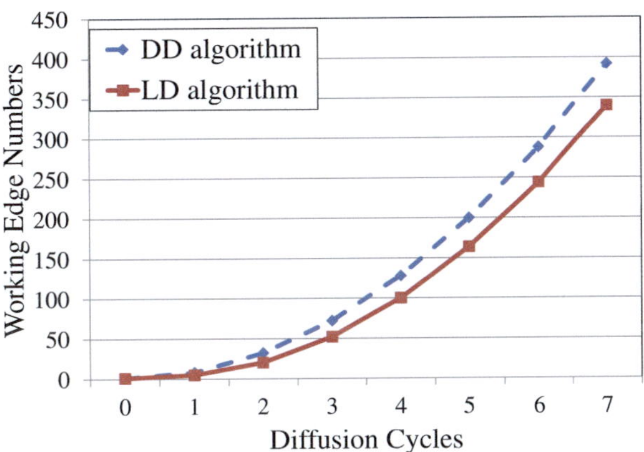

Fig. 4.4 The number of camera participants, in theory

4.4 Evaluation

We implemented three types of *Firework.Node* on A3 using three open source software: FFmpeg [20] for video decoding, OpenCV [19] for image processing and OpenALPR [18] for recognizing license plates. To demonstrate our extended version of *Firework* and to evaluate A3, we first

4.4.1 Experimental Setup

We have built a testbed for A3 consisting of 81 virtual machines on the Amazon EC2. All the machines have the same CPU (i.e., Intel Xeon CPU at 2.4 GHz) and they can communicate with each other. However, we define the application topology, to control communications. Because of Amazon EC2's limitations of service, we deploy these virtual machines in four data centers under two Amazon accounts. In this testbed, we do not consider pulling the video from cameras. We store the video data in the virtual machines and control the playing speed to simulate a live video stream.

We also built a local testbed comprised of several desktops, to evaluate the local edge nodes' collaborative performance. The reason we built two testbeds is that the cloud one is mainly for demonstrating the application and task-scheduling functions, and the local one is closer to reality.

The experimental video data are collected on a large-scale campus with 25,000 students. The resolution of all the video is 1280 × 720 pixels and 25 frames per second. The video data are encoded in H.264 format, with a baseline profile where one intra-frame is followed by 49 predictive-frames, which is typical for a live video stream.

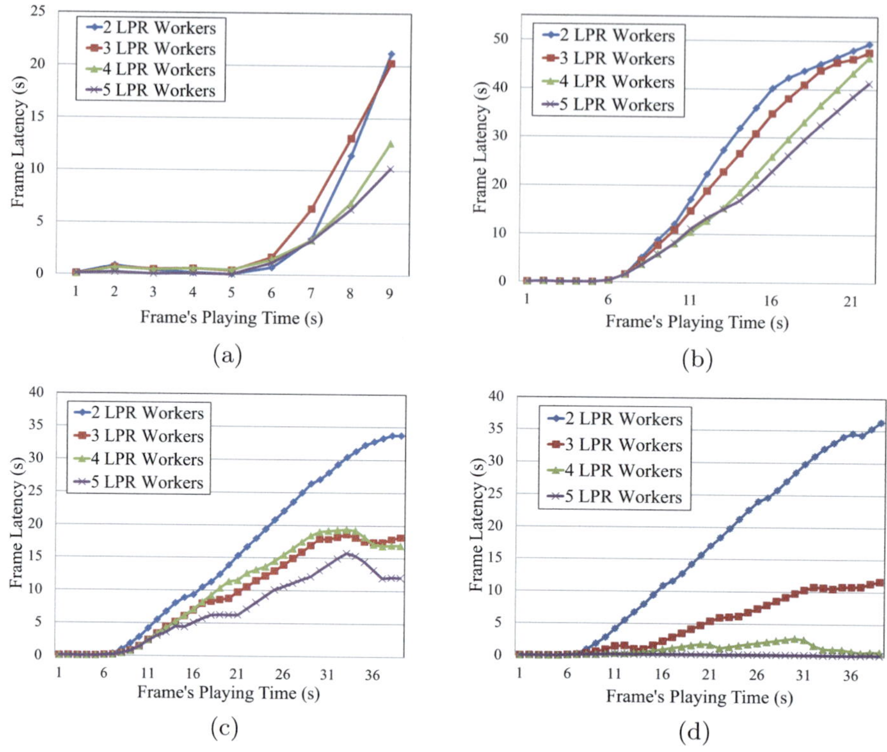

Fig. 4.5 Frame latency over time in different types of AWS nodes with different numbers of LPR workers. (**a**) AWS EC2 t2.small. (**b**) AWS EC2 t2.medium. (**c**) AWS EC2 t2.xlarge. (**d**) AWS EC2 t2.2xlarge

4.4.2   Collaboration of Local Edge Nodes

In this section, first, we set up multiple workers on one edge node, to try to achieve real-time video analytics using one edge node. The reason we do this is that video analytics benefits from parallel processing and this experiment reveals how many worker instances can process video in real time. We use four types of Amazon EC2 virtual machines: t2.small, t2.medium, t2.xlarge, and t2.2xlarge. All of them have the same CPU core (an Intel Xeon at 2.40 GHz), but each one has a different number of cores (one, two, four, and eight cores, respectively). To demonstrate local edge nodes' collaboration, we set up two edge nodes in the cloud and three edge nodes locally, to collaboratively analyze the video and evaluate the performance of these two cases regarding frame latency (defined as the time duration between when a video frame is generated and recognized). The edge nodes on the cloud are Amazon EC2 t2.xlarge, and the local edge nodes are on a Dell OptiPlex desktop with an Intel i5-4590 at 3.3 GHz.

Figure 4.5 shows the average frame latency of each second, regarding the different number of LPR worker instances implemented by a *Firework.Node*. Note

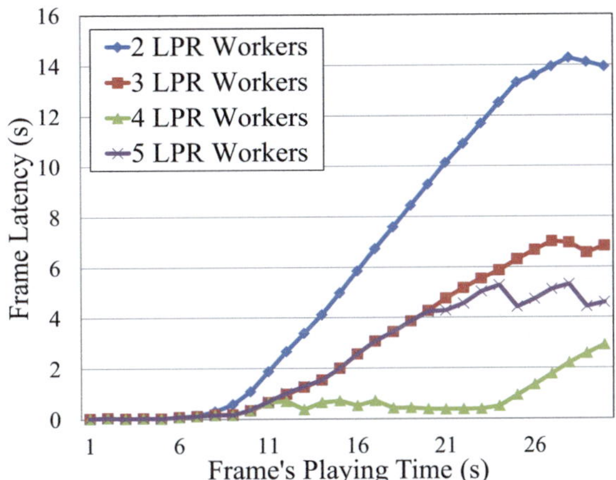

Fig. 4.6 Frame latency over time on a Dell OptiPlex with different numbers of LPR workers

that we have one video decoding worker instance and two motion detection (MD) worker instances in every *Task Receiver* for video decoding and motion detection. In general, when more LPR instances are running on the edge nodes, we see lower frame latency. According to the experiment in Sect. 4.2.1, the average processing time for plate recognition is less than 160 ms for the Amazon EC2 node. As Fig. 4.5d shows, though, the Amazon EC2 t2.2xlarge can process the video in real time with five LPR instances. However, this is only one that achieves real-time video analytics; all the others do not. The reason other Amazon EC2 nodes are difficult to process in real time because of the limited number of CPU cores. It is worth noting, though, that these nodes still achieve much lower latency—for several subsequent frames— than previous efforts (see, for example, the frames after 31 seconds). According to the CDF analysis shown in Fig. 4.1 and the processing log, the explanation for this is that there are no motion areas in some video fragments. Because it does not generate any plate-recognition workloads for plate-recognition subservice, it reduces the LPR worker's workload.

As we mentioned, we also measure the performance on local edge nodes. The results are as shown in Fig. 4.6, where for all cases it is difficult to process the video in real time, and the case including five license plate instances is worse than the case which only has four instances. This is because of the limitation of the number of CPU threads per core. The local edge node we used is a Dell OptiPlex desktop with Intel i5-4590 at 3.3GHz, which is a 4-cores and 4-threads CPU. When the number of working threads is more than the number of the CPU's threads, it will cost a lot to switch threads. Note that the CPU core's frequency shortens the processing time for each frame, and the number of the CPU's threads per core is related to how many threads run at the same time.

The experiments as mentioned earlier show that when an edge device wants to process video in real time, the CPU's threads maximum must be large enough. Generally, it should be more than four threads, but this cannot always be satisfied. Thus, *Firework* allows the *Task Dispatch* module to dispatch an overloaded task

Table 4.1 The case description for the cloud environment

	Firework.Node	MD instances' No.	LPR instances' No.
Case 1	*Task Receiver*	2	2
	Data Processor	–	1
Case 2	*Task Receiver*	2	2
	Data Processor	–	3
Case 3	*Task Receiver*	2	3
Case 4	*Task Receiver*	2	5

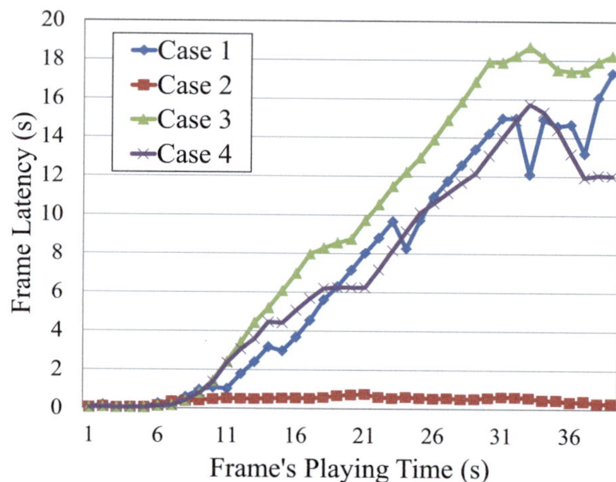

Fig. 4.7 Frame latency over time as AWS nodes collaborate

to other *Firework.Node*s, which also provides the same subservice. In A3, *Data Processor*s play this role.

According to the results of Amazon EC2 t2.xlarge in Fig. 4.5c, it still cannot process the video data in real time by itself. To demonstrate the *Firework*'s collaboration on the local edge, we set up two Amazon EC2 t2.xlarge nodes for collaboration, one of which is a *Task Receiver*, and another is a *Data Processor*. Table 4.1 shows the cases we used in the cloud. For case 1 and 2, we set different numbers of LPR instances on the *Data Processor*s. Moreover, for comparison, all LPR instances running on one edge node are permanently cased 3 comparing with case 1, in which one RP instance runs on the *Task Receiver* and other two instances run on the *Data Processor*. In a similar vein, we also set up a comparison for case 2.

Figure 4.7 illustrates the average frame latency of each second regarding all the different cases. From the results, we see that collaborative solutions are better. As we mentioned, case 4 cannot process video in real time even though it used five LPR instances (it is limited by the core number of Amazon t2.xlarge node's CPU). However, case 2 processes the video in real time, which has the same number of RP instances. Thus, a collaborative solution avoids the limitation of the number of cores in a multithread data processing application.

Table 4.2 The case description for the local environment

	Firework.Node	MD instances' No.	LPR instances' No.
Baseline	Task Receiver	2	3
Case 1	Task Receiver	2	1
	Data Processor	–	2
Case 2	Task Receiver	2	1
	Data Processor 1	–	2
	Data Processor 2	–	2

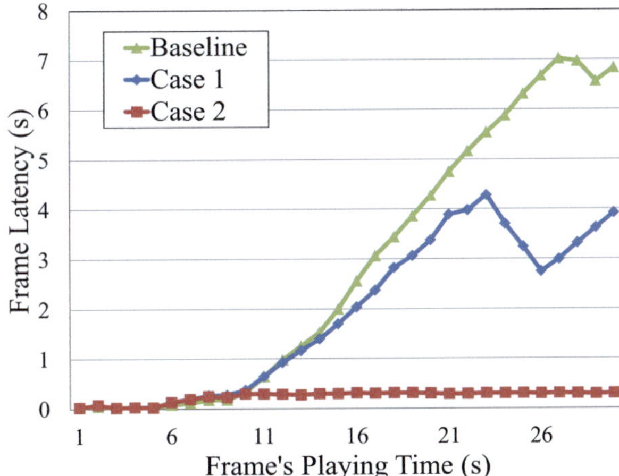

Fig. 4.8 Frame latency over time under the collaboration of local edge nodes

We also evaluate the collaborative performance using several local edge devices, which is closer to reality. Table 4.2 describes the configuration of each case we used. Because the computation resource of local edge devices is less than the Amazon EC2 t2.xlarge, we set a three-node case (see case 2 in Table 4.2) to try getting real-time video processing.

Figure 4.8 shows the results. Case 1 is better than the baseline, although they have the same number of LPR instances (but case 1 benefits from the edge nodes' collaboration). When we increase the number of *Data Processor*s to two, it can process video in real time.

4.4.3   Task Scheduling

Here, we evaluate the performance of A3's task-scheduling part on our testbed. Figure 4.3 shows the road topology we used in this experiment. We deployed 80

Task Receivers to simulate 80 edge nodes connected with road cameras. All of these *Task Receiver* are hosted on the Amazon EC2 t2.2xlarge. We deployed one *Control Center* on a low-performance Amazon EC2 virtual machine (that is, the t2.small with one core CPU and 512 MB memory). We did not deploy any *Data Processor*s on the testbed, because we apply the eight-core CPU for our virtual machine, which can analyze the video in real time using five PR worker instances. Note that the reasons we apply such powerful CPUs are multifold. First, we achieve real-time video analytics when the local edge nodes collaborate. Second, in this section, we mainly want to demonstrate the task-scheduling function. Last, one Amazon account can apply only a limited number of Amazon EC2 nodes. In this experiment, we concentrate on simulating task scheduling and evaluating the performance. Because of the Amazon EC2 service's limitations, we deployed those 81 nodes in four data centers, including Ohio, northern California, Oregon, and northern Virginia. We deployed the appropriate video data in *Task Receivers*, to make sure that four of them will locate the vehicle four times, around the 37th, 75th, 125th, and 150th seconds. In this experiment, the number of working instances is recorded per 100 milliseconds, to quantize the workloads.

We demonstrate our two task-scheduling strategies by setting the corresponding application defined topology in the *Control Center*. Then, we run the experiments several times for an average result. Figure 4.9 shows the results. Case 1 is the result of applying the DD strategy, and case 2 applies the LD strategy. The workload of all edge nodes in case 2 is less than case 1 before 200s, and once the targeted vehicle is located, the workload reduces immediately. Then, as the searching area extends again, the workload increases. Last, the workload of two cases will increase to the same value. This is because we only use 80 edge nodes to track the vehicle, and as time passes, all of them will participate in tracking the vehicle.

4.5   Related Work

Inspired by low-latency analytics, edge computing [21] (also known as fog computing [22], mobile edge computing [23], and Cloudlet [24]) processes data at the proximity of data sources. Satyanarayanan et al. [24] proposed Cloudlet, for example, which uses servers located at the edge of the network, so that computationally intensive processing can be offloaded to these edge servers. Habak et al. [25] proposed a dynamic, self-configuring, and multidevice mobile cloud out of a cluster of mobile devices, which provides a cloud service at the edge. Fernando et al. [26] also proposed a similar cloud of mobile devices. Saurez et al. [27] proposed a programming infrastructure for the distributed computational continuum represented by fog nodes and the cloud, called Foglets. *Firework* differs

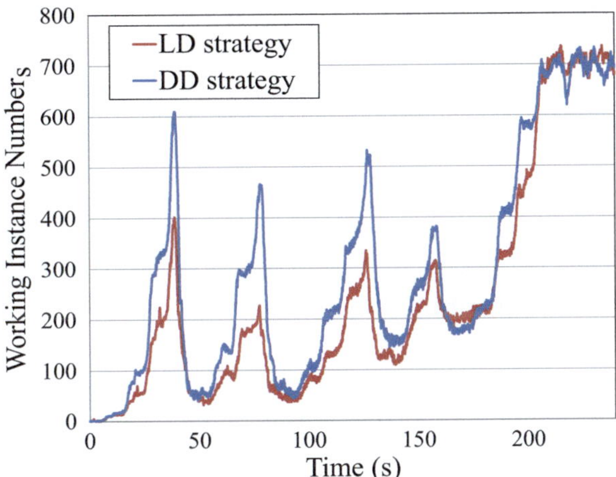

Fig. 4.9 Comparison results of two task-scheduling strategies

from each of these systems, because it leverages not only mobile devices and the cloud, but also edge nodes to complete tasks collaboratively; the other systems as mentioned earlier are not for large-scale data processing and sharing among multiple stakeholders.

As a killer application, several edge video analytics platforms have been proposed. Ananthanarayanan et al. [28] present a distributed framework for large-scale video analytics, which meets the strict requirements of the real time. It carries out different computation modules using computer vision by leveraging the public cloud, private clouds, and edge nodes. The difference between *Firework* and Ananthanarayanan's work is that our work expands data sharing, along with attached computing modules (such as functions in *Firework.View*) and programming interfaces are provided for developers to build their application on edges and the cloud. Wang et al. [29] proposed a real-time face recognition and tracking framework, called OpenFace, which also used edge computing to analyze live video. To protect privacy, OpenFace selectively blurs faces in video data, depending on user-specific privacy policies. However, OpenFace only considers edge nodes. In our work, *Firework* and A3 leverage both edge nodes and the cloud.

Zhang et al. [30] proposed a real-time video analytics platform, called VideoStorm, which leverages large clusters. It pushes all the video to clusters, and it has prohibitive costs. Our system use edge nodes to reduce costs concerning latency and network bandwidth. Yi et al. [31] proposed a latency-aware video analytics platform, called LAVEA. In LAVEA, the video data will be pushed to an edge-front node, and each video frame will be decoded and analyzed at other local edge nodes. This is similar to the collaboration of local edges. They also proposed several scheduling strategies to reduce latency. As with OpenFace, it only leverages edge nodes. Long et al. [33] proposed an edge computing framework for cooperative video processing in the IoT domain. They use mobile devices as edges to enhance the computing power and network quality by multiple uploading paths. Grassi et al. [32] proposed an application called ParkMaster, which detects parking spaces and reports empty ones to the cloud for sharing this information with other people. It uses smartphones as edge devices in the vehicle for detection. Because detection algorithms usually cost much less than the recognition algorithms, smartphones can process the video in real time.

4.6 Summary

In this chapter, we investigated the barriers of designing and implementing a distributed collaborative execution on edge, such as a real-time vehicle-tracking application that improves the *AMBER Alert* system significantly. To attack these barriers, we extended a significant data processing and sharing framework in an edge-cloud environment, to support a collaboration of local edge nodes and a customizable task-scheduling scheme. Inspired by a real project (Project Green Light in Detroit), we abstracted out the network model, applying in A3. Based on *Firework*'s extensions, we implemented the *AMBER Alert Assistant* (A3), which supports different tracking strategies. Then, we evaluated this application's performances. The results show that A3 can analyze video streams in real time by collaborating with several edge nodes, and the proposed location-direction-related task-scheduling strategy (LD) is more efficient at controlling the search area for vehicle tracking. This demonstrates that A3 is ready to deploy on top of Project Green Light, and we believe A3 will improve the *AMBER Alert*.

Currently, *Firework* does not provide an interface to adjust the deployment plan, so A3 cannot dynamically adjust the diffusion rate. This is helpful in a real scenario, considering road conditions. For future work, we will design and implement such a module. Besides, an elastic message queue module is also our future work, which can dynamically launch the best suitable message queue system for each edge device and a lightweight inner message queue for modules in *Firework*.

In the next chapter, we will summarize the challenges as well as the opportunities in Edge computing and introduce some solutions that could be employed.

References

1. M. Armbrust, A. Fox, R. Griffith, A. D. Joseph, R. Katz, A. Konwinski, G. Lee, D. Patterson, A. Rabkin, I. Stoica *et al.*, "A view of cloud computing," *Communications of the ACM*, vol. 53, no. 4, pp. 50–58, 2010.
2. S. Ghemawat, H. Gobioff, and S.-T. Leung, "The google file system," in *ACM SIGOPS operating systems review*, vol. 37, no. 5.   ACM, 2003, pp. 29–43.
3. K. Shvachko, H. Kuang, S. Radia, and R. Chansler, "The hadoop distributed file system," in *Proceedings of IEEE 26th Symposium on Mass Storage Systems and Technologies (MSST).* IEEE, 2010, pp. 1–10.

4. "Apache storm," Feb. 2017. [Online]. Available: https://storm.apache.org/

5. M. Zaharia, M. Chowdhury, M. J. Franklin, S. Shenker, and I. Stoica, "Spark: cluster computing with working sets," in *Proceedings of the 2nd USENIX Conference on Hot Topics in Cloud Computing*, vol. 10, 2010, p. 10.

6. M. Zaharia, T. Das, H. Li, T. Hunter, S. Shenker, and I. Stoica, "Discretized streams: Fault-tolerant streaming computation at scale," in *Proceedings of the 24th ACM Symposium on Operating Systems Principles*. ACM, 2013, pp. 423–438.

7. Q. Zhang, Y. Song, R. Routray, and W. Shi, "Adaptive block and batch sizing for batched stream processing system," in *Proceedings of IEEE International Conference on Autonomic Computing (ICAC)*, July 2016, pp. 35–44.

8. J. Dean and S. Ghemawat, "Mapreduce: simplified data processing on large clusters," *Communications of the ACM*, vol. 51, no. 1, pp. 107–113, 2008.

9. D. E. Culler, "The once and future internet of everything," *GetMobile: Mobile Computing and Communications*, vol. 20, no. 3, pp. 5–11, 2017.

10. "First Workshop on Video Analytics in Public Safety," https://www.nist.gov/sites/default/files/documents/2017/01/19/ir_8164.pdf, [Online; accessed Feb. 1st, 2017].

11. "Cisco global cloud index: Forecast and methodology, 2014-2019 white paper," 2014.

12. D. Evans, "The internet of things: How the next evolution of the internet is changing everything," *CISCO white paper*, vol. 1, no. 2011, pp. 1–11, 2011.

13. K. Pulli, A. Baksheev, K. Kornyakov, and V. Eruhimov, "Real-time computer vision with opencv," *Communications of the ACM*, vol. 55, no. 6, pp. 61–69, 2012.

14. Q. Zhang, X. Zhang, Q. Zhang, W. Shi, and H. Zhong, "Firework: Big data sharing and processing in collaborative edge environment," in *2016 Fourth IEEE Workshop on Hot Topics in Web Systems and Technologies (HotWeb)*, Oct 2016, pp. 20–25.

15. Q. Zhang, Q. Zhang, W. Shi, and H. Zhong, "Firework: Data processing and sharing for hybrid cloud-edge analytics," *Technical Report MIST-TR-2017-002*, 2017.

16. (2017, Mar.) Amber alert. [Online]. Available: https://en.wikipedia.org/wiki/AMBER_Alert

17. (2017, Mar.) Project green light. [Online]. Available: http://www.greenlightdetroit.org/

18. (2017, Mar.) Openalpr. [Online]. Available: https://github.com/openalpr/openalpr

19. (2017, Mar.) Opencv. [Online]. Available: http://www.opencv.org/

20. (2017, Mar.) Ffmpeg. [Online]. Available: https://ffmpeg.org/

21. W. Shi, J. Cao, Q. Zhang, Y. Li, and L. Xu, "Edge computing: Vision and challenges," *IEEE Internet of Things Journal*, vol. 3, no. 5, pp. 637–646, 2016.

22. F. Bonomi, R. Milito, J. Zhu, and S. Addepalli, "Fog computing and its role in the internet of things," in *Proceedings of the first edition of the MCC workshop on Mobile Cloud Computing*. ACM, 2012, pp. 13–16.

23. M. Patel, B. Naughton, C. Chan, N. Sprecher, S. Abeta, A. Neal *et al.*, "Mobile-edge computing introductory technical white paper," *White Paper, Mobile-edge Computing (MEC) industry initiative*, 2014.

24. M. Satyanarayanan, P. Bahl, R. Caceres, and N. Davies, "The case for vm-based cloudlets in mobile computing," *Pervasive Computing*, vol. 8, no. 4, pp. 14–23, 2009.

25. K. Habak, M. Ammar, K. A. Harras, and E. Zegura, "Femto clouds: Leveraging mobile devices to provide cloud service at the edge," in *Proceedings of the International Conference on Cloud Computing*. IEEE, 2015, pp. 9–16.

26. N. Fernando, S. W. Loke, and W. Rahayu, "Computing with nearby mobile devices: a work sharing algorithm for mobile edge-clouds," *IEEE Transaction on Cloud Computing*, 2016.

27. E. Saurez, K. Hong, D. Lillethun, U. Ramachandran, and B. Ottenwälder, "Incremental deployment and migration of geo-distributed situation awareness applications in the fog," in *Proceedings of the International Conference on Distributed and Event-based Systems*. ACM, 2016, pp. 258–269.

28. G. Ananthanarayanan, P. Bahl, P. Bodík, K. Chintalapudi, M. Philipose, L. Ravindranath, and S. Sinha, "Real-time video analytics: The killer app for edge computing," *Computer*, vol. 50, no. 10, pp. 58–67, 2017.

29. J. Wang, B. Amos, A. Das, P. Pillai, N. Sadeh, and M. Satyanarayanan, "A scalable and privacy-aware iot service for live video analytics," in *Proceedings of the 8th ACM on Multimedia Systems Conference*, ser. MMSys'17. New York, NY, USA: ACM, 2017, pp. 38–49.

30. H. Zhang, G. Ananthanarayanan, P. Bodik, M. Philipose, P. Bahl, and M. J. Freedman, "Live video analytics at scale with approximation and delay-tolerance." in *NSDI*, 2017, pp. 377–392.

31. S. Yi, Z. Hao, Q. Zhang, Q. Zhang, W. Shi, and Q. Li, "Lavea: Latency-aware video analytics on edge computing platform," in *Proceedings of 2nd ACM/IEEE Symposium on Edge Computing (SEC)*, ser. SEC'17. New York, NY, USA: ACM, 2017.

32. G. Grassi, P. Bahl, J. Kyle, and G. Pau, "Parkmaster: An in-vehicle, edge-based video analytics service for detecting open parking spaces in urban environments," in *Proceedings of 2nd ACM/IEEE Symposium on Edge Computing (SEC)*, ser. SEC'17. New York, NY, USA: ACM, 2017.

33. C. Long, Y. Cao, T. Jiang, and Q. Zhang, "Edge computing framework for cooperative video processing in multimedia iot systems," *IEEE Transactions on Multimedia*, vol. PP, no. 99, pp. 1–1, 2017.

Chapter 5
Challenges and Opportunities in Edge Computing

In this chapter, we will summarize the challenges in Edge Computing and bring forward some potential solutions and opportunities worth further research, including *programmability, naming, data abstraction, service management, privacy and security* and *optimization metrics*.

5.1 Programmability

In Cloud computing, users program their code and deploy them on the cloud. The cloud provider is in charge to decide where the computing is conducted in a cloud. Users have zero or partial knowledge of how the application runs. This is one of the benefits of Cloud computing that the infrastructure is transparent to the user. Usually, the program is written in one programming language and compiled for a certain target platform, since the program only runs in the cloud. However, in the Edge computing, computation is offloaded from the cloud, and the edge nodes are most likely heterogeneous platforms. In this case, the runtime of these nodes differ from each other, and the programmer faces huge difficulties to write an application that may be deployed in the Edge computing paradigm.

To address the programmability of Edge computing, we propose the concept of *Computing Stream* that is defined as a serial of functions/computing applied to the data along the data propagation path. The functions/computing could be entire or partial functionalities of an application, and the computing can occur anywhere on the path as long as the application defines where the computing should be conducted. The computing stream is software-defined computing flow such that data can be processed in distributed and efficient fashion on data generating devices, edge nodes, and the cloud environment. As defined in Edge computing, a lot of computing can be done at the edge instead of the centric cloud. In this case, the computing stream can help the user to determine what functions/computing should be done

© The Author(s), under exclusive license to Springer Nature Switzerland AG 2018

J. Cao et al., *Edge Computing: A Primer*, SpringerBriefs in Computer Science,

https://doi.org/10.1007/978-3-030-02083-5_5

and how the data is propagated after the computing happened at the edge. The function/computing distribution metric could be latency-driven, energy cost, TCO, and hardware/software specified limitations. The detailed cost model is discussed in Sect. 5.9. By deploying a computing stream, we expect that data is computed as close as possible to the data source, and the data transmission cost can be reduced. In a computing stream, the function can be reallocated, and the data and state along with the function should also be reallocated. Moreover, the collaboration issues, (e.g., synchronization, data/state migration, etc.) have to be addressed across multiple layers in the Edge computing paradigm.

5.2 Naming

In Edge computing, one important assumption is that the number of things is tremendously large. Atop the edge nodes, many applications are running, and each application has its structure about how the service is provided. Similar to all computer systems, the naming scheme in Edge computing is significant for programming, addressing, things identification, and data communication. However, an efficient naming mechanism for the Edge computing paradigm has not been built and standardized yet. Edge practitioners usually need to learn various communication and network protocols in order to communicate with the various things in their system. The naming scheme for Edge computing needs to handle the mobility of things, highly dynamic network topology, privacy and security protection, as well as the scalability targeting the tremendously large amount of unreliable things.

Traditional naming mechanisms such as DNS and URI (Uniform Resource Identifier) satisfy most of the current networks very well. However, they are not flexible enough to serve the dynamic Edge network since sometimes most of the things at Edge could be highly mobile and resource constrained. Moreover, for some resource-constrained things at the Edge of the network, IP based naming scheme could be too massive to support considering its complexity and overhead.

New naming mechanisms such as Named Data Networking (NDN) [1] and MobilityFirst [2] could also be applied to Edge computing. NDN provides a hierarchically structured name for content/data-centric network, and it is human-friendly for service management and provides good scalability for Edge. However, it would need the extra proxy in order to fit into other communication protocols such as BlueTooth or Zigbee, and so on. Another issue associated with NDN is security since it is tough to isolate things hardware information with service providers. MobileFirst can separate name from the network address in order to provide better mobility support, and it would be very active if applied to Edge services where things are of high mobility. Nevertheless, a global unique identification (GUID) needs to be used for naming is MobileFirst, and this is not required in related fixed information aggregation service at the Edge of the network such as the home environment. Another disadvantage of MobileFirst for Edge is the difficulty in service management since GUID is not human-friendly.

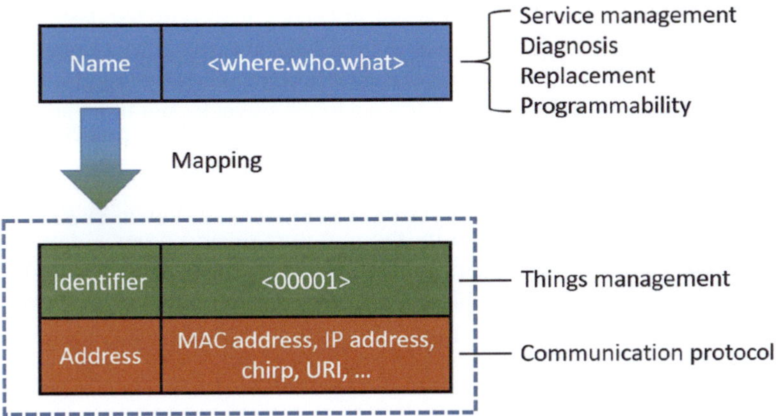

Fig. 5.1 The naming mechanism in EdgeOS

For a relatively small and fixed Edge such as home environment, let the EdgeOS assign the network address to each thing could be a solution. Within one system, each thing could have a unique human-friendly name which describes the following information: location (where), role (who), and data description (what), for example, "kitchen.oven2.temperature3". Then the EdgeOS will assign identifier and network address to this thing, as shown in Fig. 5.1. The human-friendly name is unique for each thing, and it will be used for service management, things diagnosis, and component replacement. For user and service provider, this naming mechanism makes management very easy. For example, the user will receive a message from EdgeOS like "Bulb 3 (what) of the ceiling light (who) in a living room (where) failed", and then the user can directly replace the failed bulb without searching for an error code or reconfigure the network address for the new bulb.

Moreover, this naming mechanism provides better programmability to service providers, and in the meanwhile, it blocks service providers from getting hardware information, which will protect data privacy and security better. The unique identifier and Network address could be mapped from the human-friendly name. The identifier will be used for things management in EdgeOS. Network address such as IP address or MAC address will be used to support various communication protocols such as BlueTooth, ZigBee or WiFi, and so on. When targeting highly dynamic environment such as city-level system, we think it is still an open problem and worth further investigation by the community.

5.3 Data Abstraction

Various applications can run on the EdgeOS consuming data or providing service by communicating through the APIs from the service management layer. Data abstraction has been well discussed and researched in the wireless sensor network and Cloud computing paradigm. However, in Edge computing, this issue becomes more challenging. With IoT, there would be a huge number of data generators in the network, and here we take a smart home environment as an example. In a smart home, almost all of the things will report data to the EdgeOS, not to mention a large number of things deployed all around the home. However, most of the things at the Edge of the network, only periodically report sensed data to the gateway. For example, the thermometer could report the temperature every minute, but this data will most likely only be consumed by the real user several times a day. Another example could be a security camera in the home which might keep recording and sending the video to the gateway, but the data will just be stored in the database for a certain time with nobody consuming it, and then be flushed by the latest video.

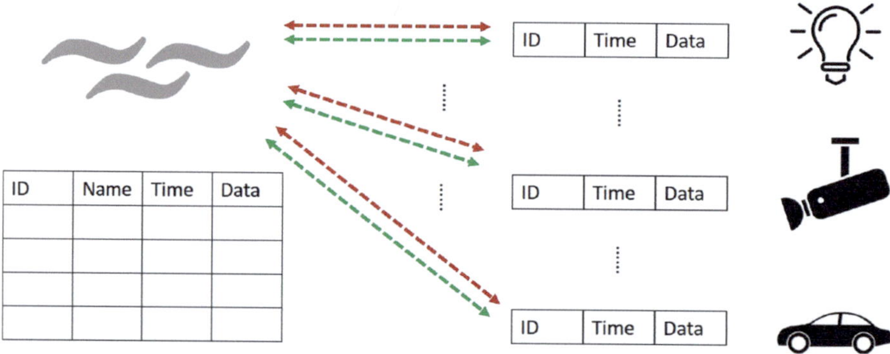

Fig. 5.2   Data abstraction issue for Edge computing

Based on this observation, we envision that human involvement in edge computing should be minimized and the Edge node should consume/process all the data and interact with users in a proactive fashion. In this case, data should be preprocessed at the gateway level, such as noise/low-quality removal, event detection, and privacy protection, and so on. Processed data will be sent to the upper layer for future service providing. There will be several challenges in this process.

First, data reported from different things comes with various formats, as shown in Fig. 5.2. For the concern of privacy and security, applications running on the gateway should be blinded by raw data. Moreover, they should extract the knowledge they are interested in from an integrated data table. We can easily define the table with id, time, name, data (e.g.,{0000, 12:34:56PM 01/01/2016, kitchen.oven2.temperature3, 78}) such that any Edge thing's data can be fitted in. However, the details of sensed data have been hidden, which may affect the usability of data.

Second, it is sometimes difficult to decide the degree of data abstraction. If too much raw data is filtered out, some applications or services could not learn enough knowledge. However, if we want to keep a large quantity of raw data, there would be a challenge for data storage. Lastly, data reported by things at Edge could be not reliable sometime, due to the low precision sensor, hazard environment, and unreliable wireless connection. In this case, how to abstract useful information from the unreliable data source is still a challenge for IoT application and system developers.

One more issue with data abstraction is the applicable operations on the things. Collecting data is to serve the application, and the application should be allowed to control, (e.g., read from and write to) the things in order to complete specific services the user desires. Combining the data representation and operations, the data abstraction layer will serve as a public interface for all things connected to EdgeOS. Furthermore, due to the heterogeneity of the things, both data representation and allowed operations could diverse a lot, which also increases the barrier of universal data abstraction.

5.4 Service Management

Concerning service management at the Edge of the network, we argue that the following four fundamental features should be supported to guarantee a reliable system, including Differentiation, Extensibility, Isolation, and Reliability (DEIR).

Differentiation With the fast growth of IoT deployment, we expected multiple services would be deployed at the Edge of the network, such as Smart Home. These services will have different priorities. For example, critical services such as things diagnosis and failure alarm should be processed earlier than ordinary service. Health-related service, for example, fall detection or heart failure detection should also have a higher priority compared with another service such as entertainment.

Extensibility Extensibility could be a considerable challenge at the Edge of the network, unlike a mobile system, the things in the IoT could be very dynamic. When the owner purchases a new thing, can it be easily added to the current service without any problem? Alternatively, when one thing is replaced due to wearing out, can the previous service adopt a new node readily? These problems should be solved with a flexible and extensible design of the service management layer in the EdgeOS.

Isolation Isolation would be another issue at the Edge of the network. In mobile OS, if an application fails or crashes, the whole system will usually crash and reboot. Alternatively, in a distributed system the shared resource could be managed with different synchronization mechanisms such as a lock or token ring. However, in a smart EdgeOS, this issue might be more complicated. There could be several applications that share the same data resource, for example, the control of light. If one application failed or was not responding, a user should still be able to control their lights, without crashing the whole EdgeOS. Alternatively, when a user removes the only application that controls lights from the system, the lights should still be alive rather than experiencing a lost connection to the EdgeOS. This challenge could be potentially solved by introducing a deployment/undeployment framework. If the OS could detect the conflict before an application is installed, then a user can be warned and avoid the potential access issue. Another side of the isolation challenge is how to isolate a user's private data from third-party applications. For example, your activity tracking application should not be able to access your electricity usage data. To solve this challenge, a well-designed control access mechanism should be added to the service management layer in the EdgeOS.

Reliability Last but not least, reliability is also a key challenge at the Edge of the network. We identify the challenges in reliability from the different views of service, system, and data here.

- From the service point of view, it is sometimes tough to identify the reason for a service failure accurately in the field. For example, if an air conditioner is not working, a potential reason could be that a power cord is cut, compressor failure or even a temperature controller has run out of battery. A sensor node could have lost connection very easily to the system due to battery outage, bad connection condition, component wear out, etc. At the Edge of the network, it is not enough to maintain a current service when some nodes lose connection but to provide the action after node failure makes more sense to the user. For example, it would be very nice if the EdgeOS could inform the user which component in the service is not responding, or even alert the user ahead if some parts in the system have a high risk of failure. Potential solutions for this challenge could be adapted from a wireless sensor network or industrial network such as PROFINET [3].
- From the system point of view, it is essential for the EdgeOS to maintain the network topology of the whole system, and each component in the system can send status/diagnosis information to the EdgeOS. With this feature, services such as failure detection, thing replacement, and data quality detection could be quickly deployed at the system level.
- From the data point of view, reliability challenge rise mostly from the data sensing and communication part. As previously researched and discussed, things at the Edge of the network could fail due to various reasons, and they could also report low fidelity data under the unreliable condition such as low battery level [4]. Also, various new communication protocols for IoT data collection are also proposed. These protocols serve well for the support of the huge number of sensor nodes and the highly dynamic network condition [5]. However, the connection reliability is not as good as BlueTooth or WiFi. If both sensing data and communication is not reliable, how can the system still provide reliable service by leveraging multiple reference data source and the historical data record is still an open challenge.

5.5   Privacy and Security

At the Edge of the network, user privacy and data security protection are the essential services that should be provided. If a home is deployed with IoT, much private information can be learned from the sensed usage data. For example, with the reading of the electricity or water usage, one can easily speculate if the house is vacant or not. In this case, how to support service without harming privacy is a challenge. Some of the private information could be removed from data before processing such as masking all the faces in the video. We think that keeping the computing at the edge of the data resource, which means in the home, could be a decent method to protect privacy and data security. To protect the data security and usage privacy at the Edge of the network, several challenges remain open.

First is the awareness of privacy and security to the community. We take WiFi networks security as an example. Among the 439 million households who use wireless connections, 49% of WiFi networks are unsecured, and 80% of households still have their routers set on default passwords. For public WiFi hotspots, 89% of them are unsecured [6]. All the stakeholders including service provider, system and application developer and end user need to aware that the users' privacy would be harmed without notice at the Edge of the network. For example, IP camera, health monitor, or even some WiFi enabled toys could easily be connected by others if not appropriately protected.

Second is the ownership of the data collected from things at Edge. Just as what happened with mobile applications, the data of end user collected by things will be stored and analyzed at the service provider side. However, leave the data at the Edge where it is collected and let the user fully own the data will be a better solution for privacy protection. Similar to the health record data, end-user data collected at the Edge of the network should be stored at the Edge, and the user should be able to control if service providers should use the data. During the process of authorization, highly private data could also be removed by the things to protect user privacy further.

Third is the missing of efficient tools to protect data privacy and security at the Edge of the network. Some of the things are highly resource constrained so the current methods for security protection might not be able to be deployed on the thing because they are resource hungry. Moreover, the highly dynamic environment at the Edge of the network also makes the network vulnerable or unprotected. For privacy protection, some platform such as Open mHealth is proposed to standardize and store health data [7], but more tools are still missing to handle different data attributes for Edge Computing.

5.6 Application Distribution

In Edge computing, the computing power of the Edge nodes is increasing, and the computing center is migrating from the Cloud to the Edge nodes. How to distribute the individual applications to various Edge nodes is one of the most important challenges, and directly related to the executable and efficiency of the Edge computing applications.

Application distribution task in Edge computing is required to decompose the applications into multiple components according to the computing resource, energy efficiency and response delay of the Edge node while keeping the semantic information of the original applications, and then distribute them to various Edge nodes. The current approaches for application distribution can be divided into two categories: dynamic and static. Static application distribution is performed during the compiling process such as Messing Passing Interface (MPI) and various multi-core application design based on GPGPU. Dynamic application distribution is performed during the running process of the applications. Similar to distributed systems, an application can be designed, implemented and debugged on center nodes so that it can run on Edge nodes. However, the Edge nodes could be heterogeneous. Thus the application distribution approaches designed for isomerism nodes are not suitable for Edge computing. The application distribution approaches for Edge computing need to consider the particularity of Cloud-Edge and Edge-Edge besides static and dynamic characters.

The idea of program dependence graph can be borrowed to determine the weight value of the heterogeneous resource, distance parameter, and communication cost between Edge nodes while adding the support for resource dependence, location dependence, and response time dependence. The analysis of the dependence can be applied to not only static and dynamic running process but also on the distribution process of Cloud-Edge and Edge-Edge relationships.

5.7 Scheduling Strategies

The scheduling strategies of Edge computing is expected to optimize the utilization of the resource, reduce the response time, improve energy efficiency, and improve the efficiency of task processing. The scheduling strategies of Edge computing also need to coordinate the computing task and resource between nodes similar to a traditional distributed system, meanwhile need to consider the heterogeneous of the computing resource similar to Cloud computing. Moreover, the computing resource of Edge computing is also constrained, unlike Cloud computing. Edge computing, how to manage the computing resource is one of the most important challenges.

The scheduling strategies of Edge computing need to be designed according to different applications based on the heterogeneous of the resource such as data, computing, storage, and network. Moreover, the strategies need to take the support of multiple types of an application into consideration. The scheduling strategies need to fully utilize the limited resource on Edge nodes to increase the executability and resource efficiency of the applications. To support the resource utilization, the application running status and dynamic resource condition need to be monitored and tracked in a real-time manner.

Current research shows that the scheduling strategies of Edge computing can be implemented based on graph theory. A graph can represent every application, and each node in the graph stands for an application component, each edge stands for the communication between different locations. The node in the graph can also represent a computing resource such as a server, and the edge stands for the relationships between resources. In this way, the application scheduling problem is transferred to the mapping problem between resource nodes and applications. During the resource scheduling processing, various components of the application can run on both center cloud as well as the resource on the edge of the network. In Edge computing, the availability of the resource, network condition, and user location is dynamic. Thus different components of an application might migrate from one node to another node.

5.8 Business Model

The business model of Cloud computing is relatively simple. Users directly purchase service based on their demand from the service provider. In detail, Cloud service is provided by Cloud computing based on the delivery model of the Internet-related service. Usually, the Cloud services are provided through dynamic, virtual, and easy to expand Internet resources. These services could be IT infrastructure, software, and other resources or services related to the Internet. The computing capability of Cloud service could also be used as the service or product, and circulate through the Internet.

Edge computing expands across multiple domains including Information Technology, Communication Technology, as well as numerous links on the industrial chain containing software/hardware platform, Internet, data aggregation, chip, sensor, and industrial applications. The business model of Edge computing should not only be drive-by user demanding services but also drive by data, as we will describe in the Firework framework. In the business model of Edge computing, the individual user will request data from the data owner, then cloud center or Edge data owner will feedback the processed result to the user. In this way, the business model is transformed from the single center–user relationship to multilateral center/user–user relationships.

Edge computing's business model depends on the stakeholders involved in the business. How to develop the multilateral Edge computing business model by combining the existing Cloud computing business model is one of the open issues to the community.

5.9 Optimization Metrics

In Edge computing, we have multiple layers with different computation capability. Workload allocation becomes a big issue. We need to decide which layer to handle the workload or how many tasks to assign at each part. There are multiple allocation strategies to complete a workload, for instances, evenly distribute the workload on each layer or complete as much as possible on each layer. The extreme cases are fully operated on the endpoint or fully operated on the cloud. To choose an optimal allocation strategy, we discuss several optimization metrics in this section, including latency, bandwidth, energy, and cost.

Latency Latency is one of the most important metrics to evaluate the performance, especially in interaction applications/services [8, 9]. Servers in Cloud computing provide high computation capability. They can handle complex workloads in a relatively short time, such as image processing, voice recognition and so on. However, latency is not only determined by computation time. Long WAN delays can dramatically influence the real-time/interaction intensive applications' behavior [10]. To reduce the latency, the workload should better be finished in the nearest layer which has enough computation capability to the things at the Edge of the network. For example, in the smart city case, we can leverage phones to process their local photos first then send a potential missing child's info back to the cloud instead of uploading all photos. Due to a large number of photos and their size, it will be much faster to pre-process at the edge. However, the nearest physical layer may not always be a good option. We need to consider the resource usage information to avoid unnecessary waiting time so that an optimal logical layer can be found. If a user is playing games, since the phone's computation resource is already occupied, it will be better to upload a photo to the nearest gateway or micro-center.

Bandwidth From latency's point of view, high bandwidth can reduce transmission time, especially for large data, (e.g., video, etc.) [11, 12]. For short distance transmission, we can establish high bandwidth wireless access to send data to the edge. On the one hand, if the workload can be handled at the edge, the latency can be significantly improved compared to work on the cloud. The bandwidth between the edge and the cloud is also saved. For example, in the smart home case, almost all the data can be handled in the home gateway through Wi-Fi or other high-speed transmission methods. Besides, the transmission reliability is also enhanced as the transmission path is short. On the other hand, although the transmission distance cannot be reduced since the edge cannot satisfy the computation demand, at least the data is pre-processed at the edge, and the upload data size will be significantly reduced. In the smart city case, it is better to pre-process photos before upload so that the data size can be significantly reduced. It saves the users' bandwidth, especially if they are using a carriers' data plan. From a global perspective, the bandwidth is saved in both situations, and it can be used by other edges to upload/download data. Hence, we need to evaluate if a high bandwidth connection is needed and which speed is suitable for an edge. Besides, to correctly determine the workload allocation in each layer, we need to consider the computation capability and bandwidth usage information in layers to avoid competition and delay.

Energy Battery is the most precious resource for things at the Edge of the network. For the endpoint layer, offloading workload to the edge can be treated as an energy-free method [13, 14]. So for a given workload, is it energy efficient to offload the whole workload (or part of it) to the edge rather than compute locally? The key is the trade-off between the computation energy consumption and transmission energy consumption. Generally speaking, we first need to consider the power characteristics of the workload. Is it computation intensive? How many resources will it use to run locally? Besides the network signal strength [15], the data size and available bandwidth will also influence the transmission energy overhead. We prefer to use Edge computing only if the transmission overhead is smaller than computing locally. However, if we care about the whole Edge computing process rather than only focus on endpoints, total energy consumption should be the accumulation of each used layer's energy cost. Similar to the endpoint layer, each layer's energy consumption can be estimated as local computation cost plus transmission cost. In this case, the optimal workload allocation strategy may change. For example, the local data center layer is busy, so the workload is continuously uploaded to the upper layer. Comparing with computing on endpoints, the multi-hop transmission may dramatically increase the overhead which causes more energy consumption.

Cost From the service providers' perspective, e.g., YouTube, Amazon, etc., Edge computing provides them less latency and energy consumption, potentially increased throughput and improved the user experience. As a result, they can earn more money for handling the same unit of workload. For example, based on most residents' interest, we can put a favorite video on the building layer edge. The city layer edge can free from this task and handle more complex work. The total throughput can be increased. The investment of the service providers is the cost to build and maintain the things in each layer. To fully utilize the local data in each layer, providers can charge users based on the data location. New cost models need to be developed to guarantee the profit of the service provider as well as the acceptability of users.

Workload allocation is not an easy task. The metrics are closely related to each other. For example, due to the energy constraints, a workload needs to be complete in the city data center layer. Comparing with the building server layer, the energy limitation inevitably affects the latency. Metrics should be given priority (or weight) for different workloads so that a reasonable allocation strategy can be selected. Besides, the cost analysis needs to be done at runtime. The interference and resource usage of concurrent workloads should be considered as well.

5.10 Summary

In this chapter, we introduced the challenges and opportunities for Edge computing, including programmability, naming, data abstraction, service management, privacy, security, and optimization metrics. In the next chapter, we will discuss the high-level literature review of the existing tools and software that could be employed for Edge computing.

References

1. L. Zhang, D. Estrin, J. Burke, V. Jacobson, J. D. Thornton, D. K. Smetters, B. Zhang, G. Tsudik, D. Massey, C. Papadopoulos *et al.*, "Named data networking (ndn) project," *Relatório Técnico NDN-0001, Xerox Palo Alto Research Center-PARC*, 2010.
2. D. Raychaudhuri, K. Nagaraja, and A. Venkataramani, "Mobilityfirst: a robust and trustworthy mobility-centric architecture for the future internet," *ACM SIGMOBILE Mobile Computing and Communications Review*, vol. 16, no. 3, pp. 2–13, 2012.
3. J. Feld, "PROFINET-scalable factory communication for all applications," in *Factory Communication Systems, 2004. Proceedings. 2004 IEEE International Workshop on*. IEEE, 2004, pp. 33–38.
4. J. Cao, L. Ren, Z. Yu, and W. Shi, "A framework for component selection in collaborative sensing application development," in *10th IEEE Conference on Collaborative Computing: Networking, Applications and Worksharing*. IEEE.
5. F. DaCosta, *Rethinking the Internet of Things: a scalable approach to connecting everything*. Apress, 2013.
6. "Wifi network security statistics/graph," http://graphs.net/wifi-stats.html/.
7. "Open mhealth platform," http://www.openmhealth.org/.
8. K. Jackson, L. Ramakrishnan, K. Muriki, S. Canon, S. Cholia, J. Shalf, H. J. Wasserman, and N. Wright, "Performance analysis of high performance computing applications on the amazon web services cloud," in *Cloud Computing Technology and Science (CloudCom), 2010 IEEE Second International Conference on*, Nov 2010, pp. 159–168.
9. A. Li, X. Yang, S. Kandula, and M. Zhang, "Cloudcmp: Comparing public cloud providers," in *Proceedings of the 10th ACM SIGCOMM Conference on Internet Measurement*, ser. IMC '10. New York, NY, USA: ACM, 2010, pp. 1–14. [Online]. Available: http://doi.acm.org/10.1145/1879141.1879143
10. M. Satyanarayanan, "Mobile computing: The next decade," *SIGMOBILE Mob. Comput. Commun. Rev.*, vol. 15, no. 2, pp. 2–10, Aug. 2011. [Online]. Available: http://doi.acm.org/10.1145/2016598.2016600
11. A. Greenberg, J. Hamilton, D. A. Maltz, and P. Patel, "The cost of a cloud: Research problems in data center networks," *SIGCOMM Comput. Commun. Rev.*, vol. 39, no. 1, pp. 68–73, Dec. 2008. [Online]. Available: http://doi.acm.org/10.1145/1496091.1496103
12. M. Armbrust, A. Fox, R. Griffith, A. D. Joseph, R. Katz, A. Konwinski, G. Lee, D. Patterson, A. Rabkin, I. Stoica, and M. Zaharia, "A view of cloud computing," *Commun. ACM*, vol. 53, no. 4, pp. 50–58, Apr. 2010. [Online]. Available: http://doi.acm.org/10.1145/1721654.1721672
13. A. P. Miettinen and J. K. Nurminen, "Energy efficiency of mobile clients in cloud computing," in *Proceedings of the 2Nd USENIX Conference on Hot Topics in Cloud Computing*, ser. HotCloud'10. Berkeley, CA, USA: USENIX Association, 2010, pp. 4–4. [Online]. Available: http://dl.acm.org/citation.cfm?id=1863103.1863107
14. B.-G. Chun, S. Ihm, P. Maniatis, M. Naik, and A. Patti, "Clonecloud: Elastic execution between mobile device and cloud," in *Proceedings of the Sixth Conference on Computer Systems*, ser. EuroSys '11. New York, NY, USA: ACM, 2011, pp. 301–314. [Online]. Available: http://doi.acm.org/10.1145/1966445.1966473
15. N. Ding, D. Wagner, X. Chen, A. Pathak, Y. C. Hu, and A. Rice, "Characterizing and modeling the impact of wireless signal strength on smartphone battery drain," *SIGMETRICS Perform. Eval. Rev.*, vol. 41, no. 1, pp. 29–40, Jun. 2013. [Online]. Available: http://doi.acm.org/10.1145/2494232.2466586

Chapter 6
Existing Edge Computing Tools

In this chapter, we will introduce a few essential tools and software enabling edge computing. What the tools and software appear in this chapter is a tip of the thousands of open-sourced or production-ready tools and software available in the community. Thus, this chapter serves as a high-level literature review of representatives of the most popular tools and software.

6.1 What Is Your Role in Edge Computing?

To help you to answer this question, Fig. 6.1 shows a high-level overview of the hardware hierarchy of edge computing paradigm. On the top of this hierarchy, the cloud still plays an important role to perform a centralized data processing/analysis, such as machine learning, AI application, and data storage. In the middle, the edge computing acts as the regional center of geo-distributed applications, where a significant portion of data processing workload is carried out on various type hardware which could be a micro datacenter, a router, or a base station. At the bottom, edge devices act as the connectors or bridges between end users and edge services or applications. Any devices with sensing capabilities that collect data from the users and surrounding environment cloud be an edge device, including a smartphone, a laptop, a dash camera on the vehicle. It is worth noting that there are overlaps between the edge servers and edge devices and there is not a clear border between them.

Given the hierarchy, as a *provider* of hardware or infrastructural/necessary capabilities for others to leverage edge computing resources, the responsibility of a *provider* will be resource virtualization, provisioning, and managing and making these resources transparent to users. On the other hand, as an edge application/service *developer*, what developers care the most is what kind of edge devices can be leveraged by their applications, what tools or data processing platforms are the best

J. Cao et al., *Edge Computing: A Primer*, SpringerBriefs in Computer Science,
https://doi.org/10.1007/978-3-030-02083-5_6

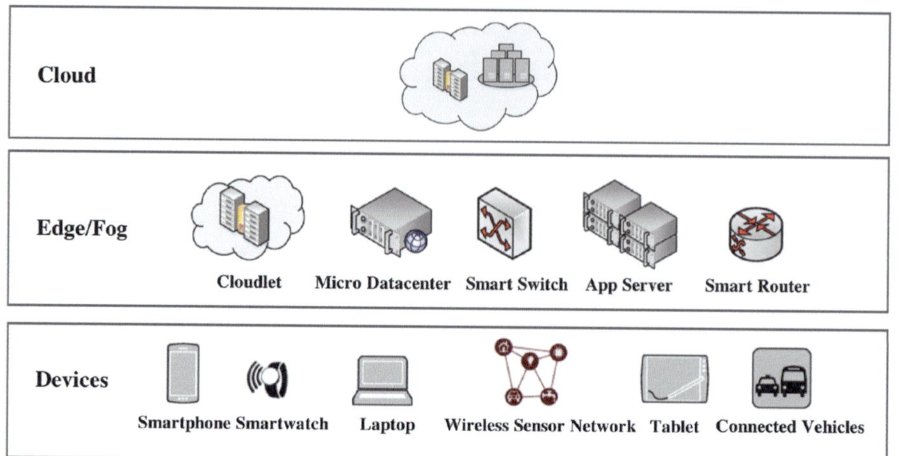

Fig. 6.1 A high-level hardware hierarchy of edge computing paradigm

fit to their applications, and how to develop, test, and deploy their applications in edge computing.

In this chapter, we categorize two significant roles in edge computing as *provider*, and *developer* and discuss what the tools and software each role can leverage to facilitate its responsibility is. In the rest of this chapter, we will first introduce the tools and software for the *provider*, including virtualization techniques for computing and networking resources, and how to efficiently manage the virtualized resources. For the *developer*, a high-level overview of the architecture for end-to-end edge analytics is presented. Tools and software that can be used to build edge intelligence service are covered. Moreover, few development kits for application developers are discussed.

6.2 Virtualization

Virtualization, by definition, is an abstraction of operating systems, computing resources, storage devices or network resources. More specifically, computation virtualization usually refers to the creation of a virtual machine that acts like a real computer with an operating system, in which applications/services hosted on these virtual machines is separated from the underlying hardware resources. With such virtualization technology, then a virtual machine (VM) packages an operating system as well as the application so that a physical server running multiple virtual machines would have different operating systems running on top of it. Similarly, the container is operating-system-level virtualization, in which the OS kernel allows the existence of multiple isolated user-space instances. In different OS, the container is also called partitions, virtualization engines or jails. An application running on an ordinary operating system can see all resources (connected devices, network shares, CPU power) of that computer. However, applications running inside a container can only see the container's contents and devices assigned to the container.

For network virtualization, it refers to a novel approach that facilitates network management and enables programmatically efficient network configuration and the abstraction of basic network functionality. Two favorite topics are the network (SDN) and network functions virtualization (NFV). SDN seeks to separate network control functions from network forwarding functions, while NFV seeks to abstract network forwarding and other networking functions from the hardware on which it runs.

The reason why virtualization is vital in edge computing is the data abstraction. An example of data abstraction is the concept of "class" in object-oriented programming, where data and operations are defined as objects. In cloud computing, data abstraction is not the first-class citizen, and all data is transferred to centralized data centers. Then different data abstraction happens in the cloud/datacenters, and different intelligent processing is applied on top of the abstracted data. However, in edge computing, the size of data is too large to prevent the data being transferred to the cloud/datacenters, which requires data abstraction happens at or near the data sources, so that data is smartly aggregated and routed to various applications, where the computation and networking virtualization can facilitate such data abstraction at the edge devices/servers.

6.2.1 Virtual Machine and Container

Virtual machine (VM) has been widely used in data centers or even standalone servers. The VM in the cloud and edge computing usually refers to the hypervisor, which emulates the underlying hardware. Based on how the hypervisor works, there are two types of hypervisor:

- Type-1, or native or bare-metal hypervisors: that run directly on the host's hardware to control the hardware and to manage guest operating systems. E.g., Hyper-V [1], Xen [2].
- Type-2, or hosted hypervisors: that run on a conventional operating system (OS), aka, host OS, and abstract guest operating systems on top of the host OS. A guest OS runs as a process on the host OS. E.g., Fusion [3], VirtualBox [4], and KVM [5].

We will not discuss the detail differences between Type-1 and Type-2 VMs, while focus more on the differences between VM and the container (which will be introduced later in this section). In this chapter, we use Type-2 VM as an example to show how the VM works. As shown in Fig. 6.2, a VM is an OS-level emulation since the software stack for a Type-2 VM includes a fully operational guest OS, runtime supports, and the application. The Fig. 6.2 indicates three VMs are hosted on the same physical machine. Type-2 hypervisors virtualize all hardware

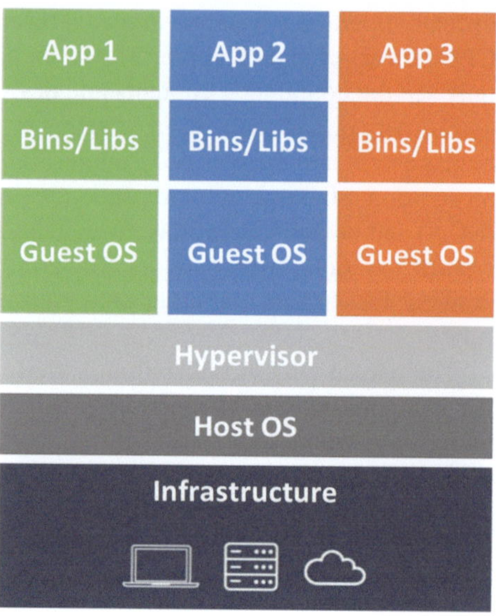

Fig. 6.2 Software stack of Type-2 hypervisor-based virtualization

and device drivers of an OS, which leads to massive overhead. However, an entirely OS emulation also allow the Type-2 hypervisor to emulate different OS on the same bare metal.

Another virtualization tool is the container, which is a resource-isolated process running an application and its dependencies. Container is built on top of two major components: (1) the Linux kernel's support for *namespaces* mostly that isolates an application's view of the operating environment, including process trees, network, user IDs and mounted file systems; and (2) the Linux kernel's *cgroups* that provide resource limiting for memory and CPU. Three examples of containers are LXC, LXD, and Docker. LXC, short for "Linux containers," is a solution for virtualizing software at the operating system level within the Linux kernel. LXD is described as a REST API that connects to libel, which is the LXC software library. Docker is similar to LXD, which is not only built on top of LXC but also integrates with container management, and we will discuss it in the resource management section.

Compared to Type-2 hypervisors, container-based virtualization provides different abstraction levels regarding virtualization and isolation. In contrast, containers avoid massive overhead by implementing process isolation at the operating system level. As shown in Fig. 6.3, there is no hypervisor and guest OS in the container-based virtualization. The container provides the os-level abstraction or process-level emulation, and a container instance combines the application with all its dependencies, without a guest OS. A container image packages a base image (which is a base OS with specific configuration), applications, dependencies (libraries, binaries), with most likely a sequence of commands to configure and start the app. The container runs as an isolated process in user space on the host OS, while the OS kernel is shared among all the containers.

Table 6.1 compares the Type-2 VM and container in six major dimensions. Since the VM emulates a fully operational OS, so the construction/destruction time is significantly longer than container whose start/stop time is just processed spawn/fork/termination. Consequently, a VM usually runs multiple services since it has full guest OS emulation, while for the container it usually hosts single service and a complex service is composed by multiple containers running one component of the service on each container. Regarding 'Guest OS,' VM supports different OS emulation on the same host OS, while the container depends on the host OS since it shares the host OS among multiple containers. As a result, a failure of VM is isolated within that VM, and another VMs on the same host OS is rarely affected. However, a failure of a container cloud causes a failure of other containers since the failure may cause host OS failure. The container has a lower migration cost, and the deployment density of container cloud be much higher than VM. For security, the container is secure, but there is no proof that the container has met the production-grade security requirements. Thus, for practice, containers and VMs can be deployed together (container on top of VM) to provide additional layers of isolation and security for selected services. There is no one-fit-all choice between VM and container, and it depends on the requirements of an application.

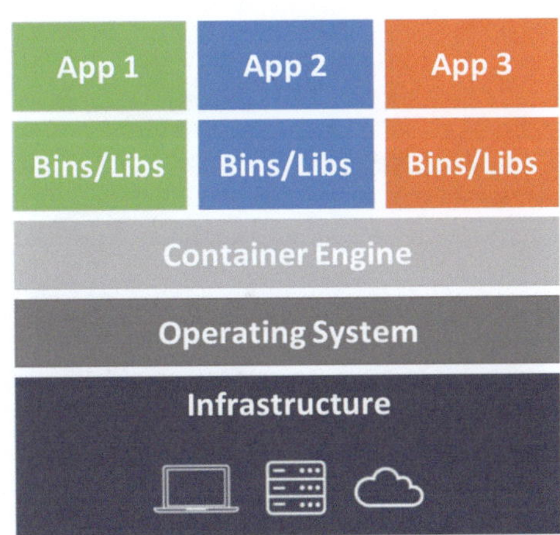

Fig. 6.3 Software stack of container based virtualization

Table 6.1 When to use VM vs container?

Factors	VM	Container
Start/stop time	10s seconds	10s milliseconds
Deployed services	Multiple services	Usually a single service
Guest OS	Independent	Depends on host OS
Failure pattern	Isolated between VMs	May affect other containers
Migration cost	More	Less but require the same host OS
Deployment density	1–10 instances	>10 instances
Security	Better	Good

6.2.2 Network Virtualization

Software-defined networking (SDN) is an architecture designed to be dynamic, manageable, and cost-effective and to be suitable for the high-bandwidth, dynamic nature of today's applications. SDN architectures decouple network control and forwarding functions, enabling network control to become directly programmable and the underlying infrastructure to be abstracted from applications and network services. Figure 6.4 shows the three layers in SDN. The control plane consists of one or more controllers which are considered as the brain of the SDN network where the whole intelligence is incorporated. The controller layer that maintains a global view of the network, which appears to applications and policy engines as a single, logical switch. However, the intelligence centralization has its drawbacks when it comes to security, scalability, and elasticity and this is the main issue of SDN. The network control (the control layer in the figure) is directly programmable by applications (application layer in the figure) because it is decoupled from forwarding functions (the data forwarding layer or infrastructure layer).

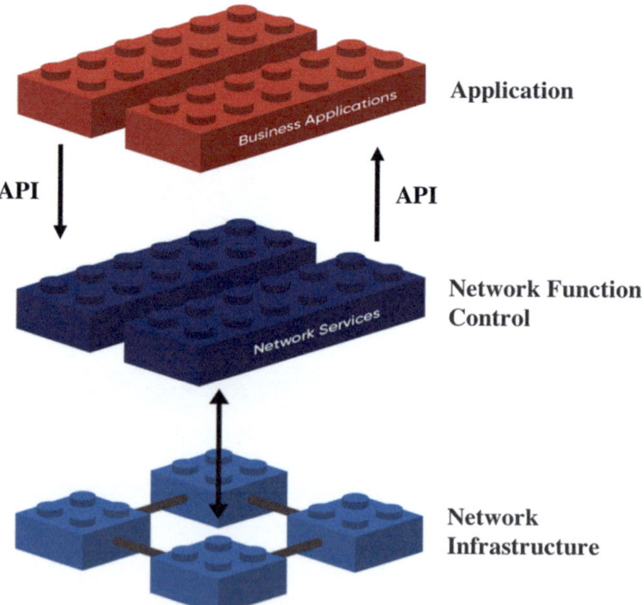

Fig. 6.4 A high-level architecture overview of SDN

Two open source projects closely related to SDN are Openflow [6] and Open vSwitch [7]. Openflow is a protocol that enables network controllers to determine the path of network packets across a network of switches. Closely related to Openflow, Open datapath is maintained and evolved with the OpenFlow protocol to expand the scope of SDN control to support a broad spectrum of hardware platforms. If you need to test your SDN application, Delta should work for you, which is an open-testing tool for SDN environments to uncover unknown security issues within an SDN deployment. Open vSwitch (OvS), which is a multi-layer virtual switch that focuses is primarily as a virtual switch, though it has been ported to various hardware platforms as well. This is also one of the goals of network function virtualization.

Network function virtualization (NFV) decouples network services from proprietary hardware appliances. As shown in Fig. 6.5, the NFV stack is similar to a VM stack mentioned in the computation virtualization section, since NFV can be treated as a dedicated VM for network functions/applications. Each network function is an individual software, which makes it easy to scale and configure, compared to traditionally dedicated appliances such as switches/routers.

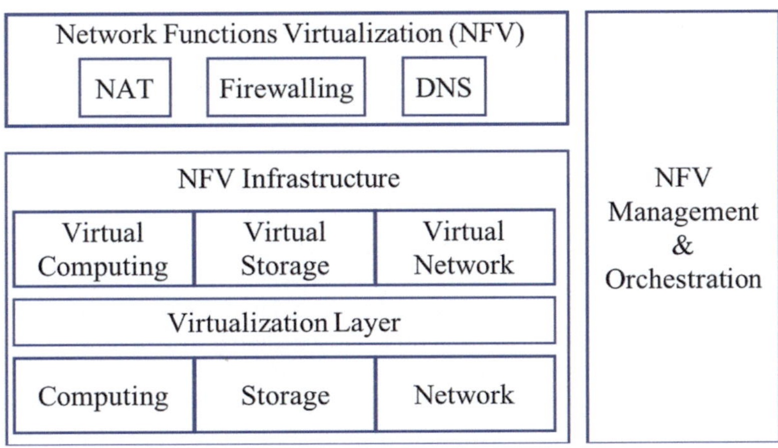

Fig. 6.5 A high-level architecture overview of NFV

Two major open source projects focusing on NFV are Doctor [8] and Pharos [9]. The doctor is fault management and maintenance framework for high availability of network services on top of the virtualized infrastructure. Doctor features immediate notification of a wide range of failure events from the NFV Infrastructure. Pharos is a federated NFV testing infrastructure of community labs around the world designed for hosting continuous integration/continuous deployment, and testing of the NFV platform, and supports orchestration for NFV recovery.

SDN and NFV differ in how they separate functions and intellectual resources. SDN abstracts physical networking resources, such as switches, routers and so on, and moves decision making to a virtual network control plane. In this approach, the control plane decides where to send traffic, while the hardware continues to direct and handle the traffic. NFV aims to virtualize all physical network resources beneath a hypervisor, which allows the network to grow without the addition of more devices. When SDN executes on an NFV infrastructure, SDN forwards data packets from one network device to another. At the same time, SDN's networking control functions for routing, policy definition, and applications run in a virtual machine somewhere on the network. Thus, NFV provides essential networking functions, while SDN controls and orchestrates them for specific uses. SDN further allows configuration and behavior to be programmatically defined and modified.

Up to this point, we have introduced few primary virtualization mechanisms and tools that can be used to virtualize edge computing resource including computing and network. Then the next issue will be how to efficiently manage these resource, given billions of edge devices and edge servers that have been and being deployed worldwide.

6.3 Resource Management

The resources in edge computing include not only traditional computing resource, like CPU, memory, storage, and network, but also some resources, such as energy, which is not the first-class citizen in cloud or grid computing. For example, the energy/power supply of a smartphone or sensor is limited, and people do not want to recharge these devices frequently, especially for the certain sensors which might be deployed for years without recharging. Thus, when implementing edge analytics using data sourced from such devices, the developers need to answer questions including how much power can be allocated for their application, and how frequently should the applications pull or push data from/to the sensors? These will be one important constraint of optimizing scheduling or workload placement. As a summary, three key concerns of the resource management in edge computing are: a high degree of heterogeneity of hardware/software, dynamic availability and mobility of hardware/software, and existing tools are relatively heavy since they are usually designed for servers and clusters.

Conventional resource management tools include operating system, VM hypervisors, dedicated resource management systems. The operating system usually is the most apparent and almost mandatory requirement for most edge devices. The operating system includes not only standard OS, such as Linux, or embedded OS for SoC but also the variant of OS which is dedicated for edge computing, such as HomeOS from Microsoft or HomeKit from Apple, which is designed for the smart home. EdgeOS_H is an on-going project that aims to connect the devices at home with the cloud, home occupants, and developers, which we have discussed in Chap. 2. For the Cloud, EdgeOS_H can upstream/downstream data and computing requests on behalf of the smart devices at home. For home occupants, EdgeOS_H provides collaboration between humans and home. For service practitioners, EdgeOS_H is also capable of reducing the complexity of development by offering a unified programming interface. VM hypervisors as we mentioned in previous sections shares resource-rich servers among multiple VM/applications. Similar to VM hypervisors, two relatively new systems, Kubernetes [10] and Docker [11] are used to coordinates multiple containers in a server or cluster. There are also dedicated resource management system such as Mesos, YARN, to allocate resources to the different application in shared clusters/cloud. The VM hypervisors and Mesos/Yarn are widely used in cloud computing, and they are already product-grade systems.

In this chapter, we will discuss more Kubernetes and Docker, which are container management system. In edge computing, most services/applications would be containerized which make it easier to development and deployment of edge analytics running on thousands of edge devices. Another reason of why container is better for edge computing is the overhead of container is relatively lighter than VM since it is process-level virtualization, which makes it more appropriate in the context of edge computing considering most devices are not resource-rich.

6.3.1 Kubernetes and Docker

Kubernetes [10] is open sourced by Google, which is a system for automating the deployment, scaling, and management of containerized applications. Figure 6.6 shows the idea of how Kubernetes manage physical resources and containers. The Kubernetes acts as a centralized scheduler to place different containers on available resources. Docker [11] is also a system for content management that provides similar capabilities as Kubernetes. In this section, we use Kubernetes as an example to discuss what are the advantages and weakness of such container management systems.

There are three major feature of Kubernetes, including

- The first feature is auto bin-packing that automatically places containers based on their resource requirements and other constraints, while not sacrificing availability, in order to drive up utilization and save even more resources. In edge computing, an edge device most likely will host multiple services/applications simultaneously. Thus it is important to smartly place different services/applications to available edge devices, so that we can achieve the best performance (e.g., achieve lower latency for VR/AR applications) if the edge devices are resource-rich, or the best energy efficiency (e.g., achieve lower data sampling frequency on certain sensors while still meet the requirement of application).
- The second feature is self-healing that is the capability to restart containers that fail, replaces and reschedules containers when nodes die, kills containers that don't respond to your user-defined health check, and doesn't advertise them to clients until they are ready to serve. In edge computing, many edge devices are mobile devices that may suffer unstable network connection issue. In the case, it is important to recover or migrate the services/applications on these devices to other resources.
- The third important feature is load balancing. Kubernetes gives containers their IP addresses and a single DNS name for a set of containers and can load-balance

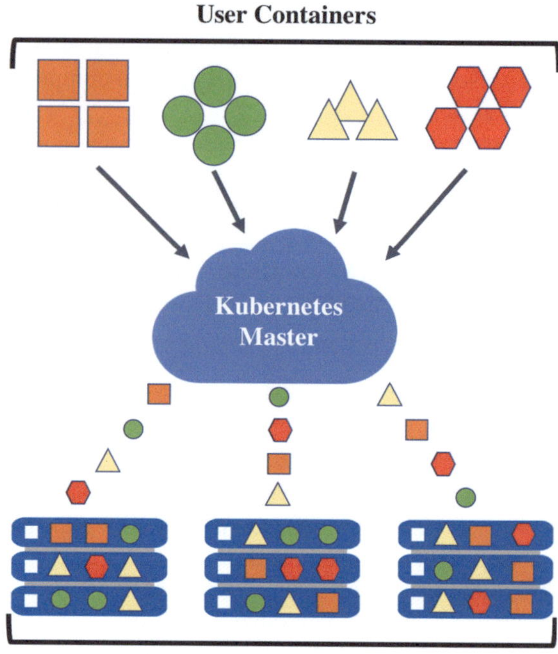

Fig. 6.6 A high-level architecture overview of Kubernetes

across them. This can also be used for on-demand workload assignment although it may violate the purpose of load-balancing.

However, Kubernets is not perfect for edge computing. One weakness of Kubernetes is that Kubernetes abstracts computing resources as a performance-equivalent unit, such as 1-CPU core equivalent to 1 vcore from AWS, GCP, or Azure. However, for heterogeneous devices which are common in edge computing, 1 CPU core differs a lot in these devices and the computing power of these heterogeneous cores cannot be guaranteed in such environment. Similarly, for memory and network resource, it is even harder to define the performance of heterogeneous devices. Another problem is Kubernetes works best for cluster management, where the hardware is usually homogeneous, and the Kubernetes has full access to all these resources. However, in edge computing, computing resources belong to different stakeholders, which make resource orchestration much harder since each stakeholder has its resource management tools, policies, and restrictions.

In summary, Kubernetes has the potential to serve as the resource manager for various edge devices, but it requires better resource abstraction and the intelligence to accommodate different resource management tools/mechanisms used by multiple stakeholders.

6.4   Developing Platforms for Edge Computing

In this section, we discuss the tools and software that could be used for service/application developers to design, prototype, and test service and applications using edge devices and cloud. First, we will illustrate an overview of the system architecture for edge analytics and what are the basic system components in typical edge analytics. Second, we will present what analytics or intelligence can edge computing carry out, instead of the cloud-based solution or complementary to the cloud-based solution. Last we will introduce several popular starter kits that can be used by developers to build their edge services/applications.

6.4.1   Edge Analytics

Edge analytics refers to data processing flow that leverage edge computing to carry out entire or partial of a complicate applications/services. Figure 6.7 shows a high-level IT architecture for edge analytics. For example, the AR/VR is very popular in video games as well as healthcare. The most critical requirement for such applications is low response latency, where the edge computing perfectly fits in the scenario by processing data at the proximity of data sources.

Another example is health care, such as fall detection for older adults, real-time vital signs monitoring (e.g., ECG, EEG). The gateway at home can collect these data and process them at home, and then if it is necessary, send an alert to a doctor. This can reduce the risk of failing to stream data from the patient to the hospital/doctor, especially when the health data is video/image based. Others like field monitoring, smart home/building/city, and autonomous robots can also leverage edge computing to improve the system performance. The Internet of Things (IoT) is the network of physical devices, including vehicles, home appliances, and other items embedded with connectivity which enables these objects to connect and exchange data. While many of today's always-connected tech devices take advantage of cloud computing, IoT application developers are starting to discover the benefits of doing more compute and analytics on the devices themselves. This on-device approach helps reduce latency for critical applications, lower dependence on the cloud, and better manage the massive deluge of data being generated by the IoT.

The edge devices cloud be sensors, cameras, or smartphones. Generally, any device that collects or generates data can contribute to edge analytics. Then these data are aggregated in nuclear devices, such as IoT gateway, on-premises server, or feed in message hubs, for local processing or partial-processing which are part of a complex application. Besides, these data sink can also redirect the same block of data to multiple downstream data consumers combining with the network switches, SDN, or NFV as discussed in previous sections. Eventually, semi-product data might be sent to the cloud, where more complex algorithms can be carried out by

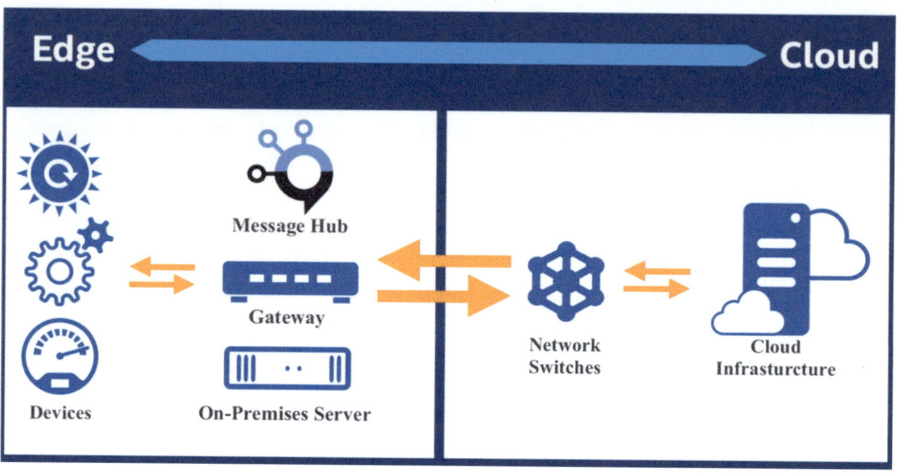

Fig. 6.7 A high-level IT architecture for edge analytics

the cloud-based big data processing system. Usually, edge devices and message hub/gateway/on-premises servers are put on the edge side. However, the border between the edge and cloud is not fixed and depends on the application, and a micro-datacenters cloud is the "cloud" of an edge device.

Most of the electrical devices will become part of edge computing, and they will play the role of data producers as well as consumers, such as air quality sensors, LED bars, street lights, and even an Internet-connected microwave oven. It is safe to infer that the number of things at the edge of the network will develop to more than billions in a few years. Thus, the raw data produced by them will be enormous, making traditional cloud computing not efficient enough to handle all these data. This means most of the data produced by edge devices will never be transmitted to the cloud. Instead, the data will be consumed at the edge of the network.

An on-premises server refers to hardware/server installed and running on the premises (in the building, the car, or on the human body) of the person or organization, rather than at a remote facility such as a server farm or cloud. In edge computing, an on-premises server may refer to a smartphone, a tablet, a PC, or dedicated appliances (e.g., Google home, Amazon Alexa/Echo) in the home environment. A microdata center/cloudlet can also server as on-premises servers for smart building/city for enterprise setup. The PC/tablet cloud be a right place for OS for edge computing/analytics, where sensors or smart appliances at home can be managed. The dedicated appliance provides many pre-installed application/services from third-parties. However, they are not programmable, and not controlled by the data owners.

Furthermore, the data owner cares more about the privacy and security issues since these devices can directly collect private data from them. Microdata center (MDC) and cloudlet usually locate at campus, organizations, buildings, etc. MDC and cloudlet are similar to the cloud environment, and they run complex analytics at the edge of the network to reduce the cost of data transmission between data sources and the cloud, and consequently provide lower response latency compared to the cloud-based solution. Thus, in edge computing, workload in the cloud is offloaded to the on-premises server and the edge devices so that data owners have full control of these devices/servers and their data.

However, some of the data collected from edge devices have to be uploaded to the cloud for complex, intelligent analysis. Given the millions of edge devices or sensors that might be used in one application, it is important to aggregate all data from these devices and feed the data to the cloud or downstream applications. Thus, the message hub serves as a bridge to queuing up the data between the edge and the cloud. There are bunch of queuing platforms, just name a few, Kafka [12], ActiveMQ [13], RabbitMQ [14]. These tools are widely used in cloud computing, and two important queuing protocols behind these queuing platforms are *advanced message queuing protocol* (AMQP) and *message queue telemetry transport* (MQTT), which can be used for edge computing with or without modification.

AMQP was designed as an open replacement for existing proprietary messaging middleware. Two of the most important reasons to use AMQP are reliability and interoperability. As the name implies, it provides a wide range of features related to messaging, including reliable queuing, topic-based publish-and-subscribe messaging, flexible routing, transactions, and security. AMQP exchanges route messages directly-in fanout form, by topic, and also based on headers. AMQP is a binary wire protocol which was designed for interoperability between different vendors. Which allows communication for edge devices from different manufacturers/vendors.

The design principles and aims of MQTT are much more straightforward and focused than AMQP. MQTT provided publish-and-subscribe messaging (no queues, despite the name) and was explicitly designed for resource-constrained devices and low bandwidth, high latency networks such as dial-up lines and satellite links, for example. It can be used effectively in embedded systems. One of the advantages MQTT has over more full-featured "enterprise messaging" brokers is that its intentionally low footprint makes it ideal for today's mobile and developing "Internet of Things" style applications, which minimize the code footprint on devices and reduce network bandwidth requirements. MQTT's strengths are simplicity (just five API methods), a compact binary packet payload (no message properties, compressed headers, much less verbose than something text-based like HTTP), and it makes a good fit for simple push messaging scenarios such as temperature updates, stock price tickers, oil pressure feeds or mobile notifications. It is also beneficial for connecting machines together, such as connecting an edge device to a web service with MQTT.

In summary, AMQP and MQTT have great potential to be used to aggregate data from millions of edge devices and based on the requirements of an application, AMQP can be used for applications requiring more reliability and interoperability to accommodate different sensors/devices from multiple vendors/data owners, while MQTT provides more lightweight data aggregation for simple data transmission.

6.4.2 Development Tools and Platforms

Processing at the edge reduces latency and makes connected applications more responsive and robust by avoiding device-to-cloud data round trips, which is critical for applications using computer vision or machine learning, for instance, an enterprise identity verification system or a drone tracking and filming its owner or an object. On-device machine learning can also enhance natural language interfaces as well, allowing smart speakers to react more quickly by interpreting voice instructions locally, such as run basic commands to switch lights on/off, or adjust thermostat settings even if internet connectivity fails. In this section, we will introduce four open source tools/platforms that are widely used for building machine learning and computer vision applications. The four tools/platforms that we discuss in the following are just a small part of available tools and platforms that can be used easily in edge computing. The developer can easily find and integrate their preferred tools and software with the mentioned tools and platforms in this section.

TensorFlow [15] is an open source software library for numerical computation using data flow graphs. The flexible architecture allows you to deploy computation to one or more CPUs or GPUs in a desktop, server, or mobile device with a single API. TensorFlow was originally developed by researchers and engineers working on the Google Brain Team within Google's Machine Intelligence research organization to conduct machine learning and deep neural networks research, but the system is general enough to be applicable in a wide variety of other domains as well. TensorFlow provides most popular machines learning algorithms including linear regression, neural network, SVM, K-means, and so on. The API is also available in multiple programming languages, such as Java, C++, Go, and Python. More importantly, it can be used in heterogeneous platforms, including CPU, GPU, as well as mobile computing platforms, and the edge devices mentioned in previous sections. This is promoting adoption of heterogeneous computer architectures, integrating different engines such as CPUs, GPUs, and DSPs, in IoT devices so that different workloads are assigned to the most efficient compute engine, thus improving performance.

OpenCV [16] was built to provide a common infrastructure for computer vision applications and to accelerate the use of machine perception in the commercial products. OpenCV has more than 2500 optimized algorithms, which includes a comprehensive set of both classic and state-of-the-art computer vision and machine learning algorithms. These algorithms can be used to detect and recognize faces, identify objects, classify human actions in videos, track camera movements, track moving objects, extract 3D models of objects, produce 3D point clouds from stereo cameras, stitch images together to produce a high resolution image of an entire scene, find similar images from an image database, remove red eyes from images taken using flash, follow eye movements, recognize scenery and establish markers to overlay it with augmented reality, etc. OpenCV provides APIs in Java, C/C++, and Python, and can be used in different computing platforms. In the era of edge computing, edge devices are being created with increasingly sophisticated computer capabilities, and edge analytics harness the benefits of the available distributed computing capabilities of field devices, gateways, and cloud altogether, to deliver faster and more robust services and intelligence.

Apache Edgent [17], previously known as Apache Quarks, is an Apache incubator project. Edgent is a programming model and micro-kernel style runtime that can be embedded in gateways and small footprint edge devices enabling local, real-time, analytics on the continuous streams of data coming from equipment, vehicles, systems, appliances, devices, and sensors of all kinds (for example, Raspberry Pis or smartphones). Working in conjunction with centralized analytic systems, Apache Edgent provides efficient and timely analytics across the whole IoT ecosystem: from the center to the edge. Apache Edgent is a streaming programming model and runtime designed to accelerate the development of analytics on small edge devices or gateways. Enabling streaming analytics on edge devices can reduce the load on data center and communication costs.

Furthermore, edge analytics can reduce decision making latency and enable device autonomy. Edgent uses a functional API to compose processing pipelines with direct per-tuple computation, and also provides a suite of connectors which significantly ease communication with a backend. The API, for now, is only available in Java.

AWS Greengrass [18] is software that lets you run the local computer, messaging, data caching, sync, and ML inference capabilities for connected devices in a secure way. AWS Greengrass seamlessly extends AWS to devices so they can act locally on the data they generate, while still using the cloud for management, analytics, and durable storage. The AWS Greengrass respond to local events in near real-time since it uses edge devices to processing the incoming events. AWS Greengrass devices can act locally on the data they generate so they can respond quickly to local events, while still using the cloud for management, analytics, and durable storage. The local resource access feature allows Lambda functions deployed on Greengrass Core devices to use local device resources like cameras, serial ports, or GPUs so that device applications can quickly access and process local data. With Amazon Lambda, you can run code for virtually any type of application or backend service—all with zero administration. Just upload your code and Lambda takes care of everything required to run and scale your code with high availability. You can set up your code to automatically trigger from other events or call it directly from any web or mobile app.

Regarding programmability, AWS Greengrass uses a simplified device programming with AWS Lambda support. AWS Greengrass uses the same AWS Lambda programming model you use in the cloud, so you can develop code in the cloud and then deploy it seamlessly to your devices. Besides, AWS Greengrass uses secure communication to protect the data privacy and security. AWS Greengrass authenticates and encrypts device data for both local and cloud communications so that data is never exchanged between devices and the cloud without proven identity.

Other similar tools and platform include but are not limited to, Deeplearning4j [19], Caffe [20], Theano [21], Torch [22], OpenNN [23] for machine learning and computing vision, and Amazon IoT [24], Microsoft Azure IoT [25], IBM Watson

IoT [26], Google Cloud IoT [27] for IoT and edge computing. The details about these tools and platforms can be easily found online, and we encourage the readers to figure out the technical details on their own and integrate them into their edge applications and services.

6.5 Summary

In this chapter, two major topics are discussed, the tools and software for resource providers and tools and software for application developers. For the resource providers, the virtualization mechanisms for computation and networking are discussed, since virtualization helps to share data/service at the edge, and container-ization simplifies resource management on heterogeneous and resource-constrained edge devices, as well as the corresponding resource management tools, especially for the containers. For the application developers, we have introduced the edge devices, messaging protocols, and developing tools and platforms that developers can use to build their application. In summary, edge computing, as a complement of the cloud, seamlessly bridges cloud-based machine intelligence with the local data.

References

1. "Microsoft Hyper-V," https://en.wikipedia.org/wiki/Hyper-V/, [Online; accessed April 10th, 2018].
2. "Xen Project," https://www.xenproject.org/, [Online; accessed April 10th, 2018].
3. "VMWare Fusion," https://www.vmware.com/products/fusion.html/, [Online; accessed April 10th, 2018].
4. "VirtualBox," https://www.virtualbox.org/, [Online; accessed April 10th, 2018].
5. "Linux KVM," https://www.linux-kvm.org/page/Main_Page/, [Online; accessed April 10th, 2018].
6. N. McKeown, T. Anderson, H. Balakrishnan, G. Parulkar, L. Peterson, J. Rexford, S. Shenker, and J. Turner, "Openflow: Enabling innovation in campus networks," *SIGCOMM Comput. Commun. Rev.*, vol. 38, no. 2, pp. 69–74, Mar. 2008. [Online]. Available: http://doi.acm.org/10.1145/1355734.1355746
7. "Open vSwitch," https://www.openvswitch.org/, [Online; accessed April 10th, 2018].
8. "Doctor," http://www.doctor-project.org/, [Online; accessed April 10th, 2018].
9. "Pharos," https://www.opnfv.org/community/projects/pharos, [Online; accessed April 10th, 2018].
10. "Kubernetes," https://kubernetes.io/, [Online; accessed April 10th, 2018].
11. (2017, Mar.) Docker. [Online]. Available: https://www.docker.com/
12. J. Kreps, N. Narkhede, and J. Rao, "Kafka: A distributed messaging system for log processing," in *Proceedings of the NetDB*, 2011, pp. 1–7.
13. "ActiveMQ," http://activemq.apache.org/, [Online; accessed April 10th, 2018].
14. "RabbitMQ," https://www.rabbitmq.com/, [Online; accessed Dec. 1st, 2016].
15. "TensorFlow," https://www.tensorflow.org/, [Online; accessed April 10th, 2018].
16. (2017, Mar.) Opencv. [Online]. Available: http://www.opencv.org/
17. "Apache edgent," http://edgent.apache.org/.

18. "Aws greengrass," https://aws.amazon.com/greengrass/.
19. "DL4J," https://deeplearning4j.org/, [Online; accessed April 10th, 2018].
20. "Caffe," http://caffe.berkeleyvision.org/, [Online; accessed April 10th, 2018].
21. "Theano," http://www.deeplearning.net/software/theano/, [Online; accessed April 10th, 2018].
22. "Torch," http://torch.ch/, [Online; accessed April 10th, 2018].
23. "OpenNN," http://www.opennn.net/, [Online; accessed April 10th, 2018].
24. "AWS IoT," https://aws.amazon.com/iot/, [Online; accessed Sep. 1st, 2016].
25. "Microsoft Azure IoT," https://www.microsoft.com/en-us/internet-of-things/, [Online; accessed April 10th, 2018].
26. "IBM Waston IoT," https://www.ibm.com/internet-of-things, [Online; accessed April 10th, 2018].
27. "Google cloud platform: IoT solution," https://cloud.google.com/solutions/iot/, [Online; accessed Sep. 1st, 2016].

Chapter 7
Conclusions

Edge computing could scale from a single person to a smart home to even an entire city. Given that a city with 1 million people will produce an estimated 180 petabytes of data per day by 2019, the benefits could be enormous.

However, to realize this vision, the systems, network, and application communities must work together, joined by the many groups that could benefit from the technology such as those in environmental and public health, law enforcement, fire protection, moreover, utility services.

In the past few years, this process has begun. For example, proponents formed the OpenFog Consortium (www.openfogconsortium.org) in November 2015 to promote an ecosystem to accelerate the adoption of open fog computing by bringing together companies, universities, and individual researchers. The OpenFog Consortium focus on collaborating with companies, research institutes, and individual researchers to accelerate the deployment of Fog computing and the development of Fog computing ecosystem. NSF and NIST listed Edge computing in their grant proposal guide in 2016. Besides, new conferences are planned, such as the IEEE/ACM Symposium on Edge Computing (SEC), held in October 2016; and the Mobile Edge Computing Congress (MEC), held in September 2016. In November 2016, Edge Computing Consortium (www.ecconsortium.net) is formed to define Edge computing architecture and standard, build Edge computing collaboration platform, push open collaboration between operation technology and information and communications technology, especially in the domain of public health, smart manufacturing, smart grid, smart home and smart driving. In 2017 we have seen companies from traditional industry began to add Edge computing into their development plan. Cloud service provider such as Amazon, Microsoft, and IBM all defined Edge computing as their next development direction. Communication service provider such as China Mobile, AT&T, Vodanfance, and Telefornica are looking for solutions fusing 5G network and Edge computing technologies to satisfy the needs of real-time, large data-oriented applications. If this trend continues, edge computing will be on its way to fulfilling its promise.

J. Cao et al., *Edge Computing: A Primer*, SpringerBriefs in Computer Science,
https://doi.org/10.1007/978-3-030-02083-5_7

Nowadays, more and more services are pushed from the cloud to the edge of the network because processing data at the edge can ensure shorter response time and better reliability. Moreover, bandwidth could also be saved if a more substantial portion of data could be handled at the edge rather than uploaded to the cloud. The burgeoning of IoT and the universalized mobile devices changed the role of the edge in the computing paradigm from the data consumer to data producer/consumer. It would be more efficient to process or message data at the edge of the network. In this brief, we came up with our understanding of Edge computing, with the rationale that computing should happen at the proximity of data sources. Then we put forward the challenges and opportunities that are worth working on, including programmability, naming, data abstraction, service management, privacy and security, business model, as well as optimization metrics. Several case studies like a smart home operating system, video analytics, hybrid Cloud-Edge analytics, and using Edge computing to enable smart firefighting are introduced to explain Edge computing in a detailed manner further.

We hope this brief will bring Edge computing to the attention of the industrial and academic community, as well as attract more computer science and engineering practitioners to work on this domain. As a new research direction, much work can be done at the vision level. However, to promote the burgeoning of Edge computing, except experts from computing systems, communications, network, and application, more researchers and institutions from related industry such as public health, environment, and service should also be deeply involved. Edge computing is a new computing model which combines multiple resources, opportunities arise in solving real-world problems and seeking new computer science research issues.